Contributions to Environmental Sciences & Innovative Business Technology

Contributions to Environmental Sciences & Innovative Business Technology (CESIBT) is an interdisciplinary series of peer-reviewed books dedicated to addressing emerging research trends relevant to the interplay between Environmental Sciences, Innovation, and Business Technology in their broadest sense. This series constitutes a comprehensive up-to-date interdisciplinary reference that develops integrated concepts for sustainability and discusses the emerging trends and practices that will define the future of these disciplines.

This series publishes the latest developments and research in the various areas of Environmental Sciences, Innovation, and Business Technology, combined with scientific quality and timeliness. It encompasses the theoretical, practical, and methodological aspects of all branches of these scientific disciplines embedded in the fields of Environmental Sciences, Innovation, and Business Technology.

The series also draws on the best research papers from EuroMid Academy of Business and Technology (EMABT) and other international conferences to foster the creation and development of sustainable solutions for local and international organizations worldwide. By including interdisciplinary contributions, this series introduces innovative tools that can best support and shape both the economical and sustainability agenda for the welfare of all countries, through better use of data, a more effective organization, and global, local, and individual work. The series can also present new case studies in real-world settings offering solid examples of recent innovations and business technology with special consideration for resolving environmental issues in different regions of the world.

The series can be beneficial to researchers, instructors, practitioners, consultants, and industrial experts, in addition to governments from around the world. Published in collaboration with EMABT, the Springer CESIBT series will bring together the latest research that addresses key challenges and issues in the domain of Environmental Sciences & Innovative Business Technology for sustainable development.

Srikanth Pulipeti • Adarsh Kumar
Nagaraju Mysore • Cathryn Peoples

Editors

Quantum and Blockchain-based Next Generation Sustainable Computing

 Springer

Editors
Srikanth Pulipeti
Mukesh Patel School of Technology
Management and Engineering
SVKM's Narsee Monjee Institute of
Management Studies Deemed-to-be
University, Shirpur Campus
Dhule, Maharashtra, India

Nagaraju Mysore
Department of Computer Science and
Engineering, GST
GITAM University
Bangalore, Karnataka, India

Adarsh Kumar
School of Computer Science
University of Petroleum and Energy Studies
Dehradun, Uttarakhand, India

School of Computer Science
Mohammed VI Polytechnic University
Benguerir, Morocco

Cathryn Peoples
School of Computing
Ulster University
Belfast, UK

ISSN 2731-8303   ISSN 2731-8311   (electronic)
Contributions to Environmental Sciences & Innovative Business Technology
ISBN 978-3-031-58070-3   ISBN 978-3-031-58068-0   (eBook)
https://doi.org/10.1007/978-3-031-58068-0

This Springer imprint is published by the registered company Springer Nature Switzerland AG
The registered company address is: Gewerbestrasse 11, 6330 Cham, Switzerland

If disposing of this product, please recycle the paper.

Contents

QSB: Smart Contracts, Consensus, and Quantum Cryptography 1
Saurabh Jain

The Intersection of Blockchain Technology and the Quantum Era for Sustainable Medical Services 19
Dinesh Kumar Atal, Vishal Tiwari, Anjali, and Rajiv Kumar Berwer

Innovative Solutions for Sustainability: Quantum and Blockchain Technologies 47
Ahmed Mateen Buttar, Nouman Arshad, and Muhammad Azeem Akbar

Combating Counterfeit Drugs in Pharmaceutical Supply Chain (PSC) Using Hyperledger Fabric Blockchain 75
Anitha Premkumar and Rajesh Natarajan

Quantum and Blockchain for Sustainable Healthcare Ecosystem 105
Syed Muzammil Munawar, Dhandayuthabani Rajendiran, and Khaleel Basha Sabjan

Music DApp on the Solana Blockchain Platform: Design, Development, and Analysis 121
Urmila Pilania, Manoj Kumar, Rohit Tanwar, Pulkit Upadhyay, and Priyanka Narayan

The Role of Blockchain in Healthcare 135
Radhika Sreedharan

Innovative Solutions for Sustainable Medical Services: A Look into Quantum and Blockchain Technologies 173
Reena Thakur, Parul Bhanarkar, Uma Patel Thakur, and Mustapha Hedabou

Blockchain Technology and Quantum Computing: A Promising Solution for the Healthcare Industry and COVID-19 Pandemic 203
Galiveeti Poornima, Deepak S. Sakkari, P. Karthikeyan, T. N. Manjunath, and K. Saritha

Sustainable Solutions for Serverless Edge, Fog, and Cloud Computing Using Quantum and Blockchain Technology 219
Sarthika Dutt, Vansh, Garima Pahwa, Aishvi Guleria, Kamya Varshney, and Deeksha Joshi

QSB: Smart Contracts, Consensus, and Quantum Cryptography

Saurabh Jain

Abstract The term "quantum-secured blockchain" pertains to a sort of blockchain technology designed to withstand attacks from quantum computers. Quantum computers are a relatively modern computers meant to work with the principles of quantum physics. They can perform certain computations substantially quicker than standard systems. One of the most significant worries regarding quantum computers is the possibility that these machines would be able to break down the encryption that has been employed to safeguard blockchain networks. Blockchains that are quantum-secured adopt strategies and mechanisms that are quantum-resistant to prevent quantum computers from accessing the data or information that is stored on the network. Even though this blockchain technology is only in its infancy stage now, it is anticipated that it will be one of the future possibilities for protecting data on blockchain networks. Investigation of new forms of encryption that cannot be broken by quantum computers, such as those based on lattices, codes, and multivariate quadratic equations, is underway.

Keywords Blockchain · Smart contract · Consensus · Digital signature · Sustainability · Quantum computing · Hashing · Quantum smart contracts

1 Introduction

The term "quantum blockchain" refers to decentralized databases that are encrypted, based on quantum computation and theory, and distributed throughout a network. Data saved in a quantum blockchain cannot be purposefully changed once it has been saved. A growing number of scholars have increasing interest on quantum blockchain research in the last few years as quantum computation and quantum

S. Jain (✉)
School of Computer Science, University of Petroleum and Energy Studies, Dehradun, India
e-mail: saurabh.jain@ddn.upes.ac.in

© The Author(s), under exclusive license to Springer Nature Switzerland AG 2024
S. Pulipeti et al. (eds.), *Quantum and Blockchain-based Next Generation Sustainable Computing*, Contributions to Environmental Sciences & Innovative Business Technology, https://doi.org/10.1007/978-3-031-58068-0_1

1

information theory have evolved. It briefly discusses the advantages of quantum blockchain over conventional blockchains and examines recent developments in the field [1]. A decentralized database called blockchain is encrypted to thwart manipulation. While blockchain technology shows great promise for various uses, current blockchain-based platforms rely on outdated encryption, hashing, and digital signature algorithms that are susceptible to attacks from quantum workstations. Individuals who have gained access to quantum computing will have an unfair benefit when it comes to mining incentives because the same considerations apply to the cryptographic hash functions or consensus that are leveraged to spawn new blocks.

Blockchain technology quickly establishes itself as a key technology that can address many global space industry issues. The smart contracts and blockchain technology have the potential to open up a plethora of new prospects for the global space industry. These include the creation of satellite payloads, supply chain management in space, and satellite as a service. Additionally, it is anticipated that blockchain will greatly impact subsequent financial transactions. The Bitcoin blockchain can be sent by satellite, which will be faster and more reliable, speeding up international financial transactions. Without using the Internet, the Blockstream Satellite Network demonstrates how satellites using the blockchain can process and send Bitcoin around the globe [2, 3].

Scalability, effectiveness, and longevity are three significant challenges for blockchain technology. These obstacles need to be overcome if blockchain is ever going to develop into a mature technology that can be used responsibly. It is possible that quantum computers would have been invented before the broad implementation of blockchain technology for mission-critical tasks in financial and other types of organizations. Quantum computing can be utilized to solve some of the challenges that arise with the implementation of certain blockchain technologies, such as cryptocurrencies. This can facilitate the installation of such technologies. This article highlights the research that has already been undertaken on hybrid quantum-classical blockchain technology, as well as the unanswered research questions [4].

Even though quantum computing, a thriving technology, poses a threat to some of the fundamental elements of blockchain technology, it is widely recognized as a significant future technology. According to current estimates [5, 6], there is a 15% chance that quantum computers will be commercially available by 2026 and a 50% chance that they will be by 2031. Since most existing blockchain systems rely mostly on digital signatures with public keys and are used to communicate value, they are especially a little bit susceptible to attacks from quantum workstations. Fedorov et al. [7] say that blockchain technology as it is now might not work if quantum technologies are not added. Existing temporary solutions, like post-quantum cryptography [8–10], don't guarantee solutions to the threat that are completely safe. There have been many studies [11–13] on the quantum-safe blockchain, which could protect against attacks from future quantum computers.

2 Blockchain and Quantum Computing

A blockchain is a type of technology that enables the formation of distributed ledgers that are not centralized and may be used to record transactions that take place across a computer network. On the contrary side, quantum computing is a subfield of computer science that tries to capitalize on the power of quantum mechanics to tackle difficulties that are now intractable with traditional computers. At the intersection of these two technological spheres, significant research and development work are currently being carried out. One possible use of this intersection is quantum-resistant cryptography, which leverages the principles of quantum physics to design cryptographic algorithms and methods resistant to quantum attacks. One further possible use case involves implementing blockchain networks with quantum computing to boost their operational effectiveness and scalability. Although it is not yet evident how these technologies will be combined or what specific applications will emerge, this is undeniably an interesting field of research [14, 15].

2.1 *Quantum Computing Effects on the Blockchain Technology*

The technology behind blockchain could be impacted in several different ways by quantum computing. Developing encryption algorithms that are more resistant to attacks from quantum algorithms is one approach that could be taken. If this were successful, blockchain networks would be safe from harm. As a consequence of the circumstance that quantum-based computers can defeat the encryption methods that are now used to protect blockchain systems, new cryptographic approaches that are immune to quantum attacks must be developed. Computing on the quantum level could also have an impact on blockchain technology by improving the scalability and efficiency of blockchain networks [15]. It is possible that the use of quantum computers, which can complete certain jobs significantly more quickly than conventional computers, could be put to use to improve the pace at which blockchain networks function. This might make it possible to process more transactions per second while at the same time reducing the amount of energy needed to keep the network secure. It should be emphasized that even though quantum computing has the prospective to affect blockchain technology, the arena of study is still in its infancy, many of the proposed applications are still in the research phase, and the nature of the effect that quantum computing will have on blockchain technology is not yet known [16, 17].

2.2   Quantum Blockchain

The convergence of blockchain technology and quantum computing gives rise to a captivating concept referred to as the quantum blockchain. The main goal of a quantum blockchain is to establish a decentralized and distributed ledger that exhibits resilience against potential attacks from quantum workstations. The quantum-resistant encryption algorithm serves to enhance the safety of the network and bolsters its resilience against potential quantum-based attacks. One of the main challenges confronted in the realm of quantum blockchain technology pertains to the scarcity and nascent state of quantum computers. However, due to the anticipated developments and increased accessibility of quantum computers, there will be a considerable increase of interest for quantum-resistant blockchains.

Furthermore, the utilization of quantum blockchain likely addresses scalability concerns and enhances the efficacy of traditional blockchain systems. Quantum algorithms possess the capacity to enhance the efficiency of transaction processing. Quantum entanglement has the potential as a viable solution for enhancing the confidentiality and integrity of blockchain transactions. It is imperative to acknowledge that the existing level of research in this domain is in its nascent stage, and the verifiable feasibility of the purported advantages of quantum blockchain remains uncertain. Furthermore, the amalgamation of quantum computing inside the framework of blockchain technology may require a substantial duration to evolve into a viable and feasible solution [18, 19].

2.3   Types of Quantum Blockchain

Multiple iterations of quantum blockchains have been suggested, each possessing distinct merits and drawbacks. This section presents a succinct outline of the prevailing categorizations of quantum blockchains, as reported in scholarly literature [20–22], and the summary is provided in Table 1.

- The quantum-secured blockchain utilizes cryptographic techniques that are resistant to quantum computer assaults to safeguard the network. The use of this strategy guarantees the persistent safety of blockchain systems, notwithstanding the escalating computational competences of quantum computers.
- Quantum-enhanced blockchain pertains to a classification of blockchain systems that utilize quantum algorithms and quantum computation to augment the scalability and performance of these networks. This has the potential to lead to an augmentation in transaction processing capacity and a decrease in the energy consumption associated with network security.
- The utilization of quantum entanglement in the quantum-federated blockchain system serves to augment the levels of privacy and security associated with transactions conducted on blockchain platforms. In a blockchain network, it is possi-

Table 1 Summary of different types of quantum blockchains [20–22]

Type	Description	Key features
Quantum-secured blockchain	The network employs quantum-resistant cryptography as a means of safeguarding against potential assaults from quantum computers	Offers more security steps to protect blockchain technology from quantum attacks. Ensures the reliability of data in the setting of developments in quantum computing. This project lays the groundwork for blockchain technology to be profitable in the long run
Quantum-enhanced blockchain	Quantum algorithms and quantum processing are utilized to enhance scalability and efficiency, with the potential to increase transaction throughput while decreasing energy consumption	Using this technology makes the transaction handling go faster. Using this technology cuts down on the amount of energy needed for mining and consensus processes. The network is now able to grow and respond better
Quantum-federated blockchain	Quantum entanglement is employed to enhance the safety and security of financial transactions. Enables several individuals to utilize a shared blockchain infrastructure without necessitating the disclosure of their respective transactional activities to one another	Transactions are kept private because strict security steps are put in place. Allows more than one organization to use a safe blockchain system. Make sure that privacy stays the same in consortium blockchain settings
Quantum-hybrid blockchain	When the benefits of quantum-secured and quantum-enhanced blockchains are combined, the result is a system that is both reliable and safe	This system has both protection and performance features that work well together. Shows the ability to deal with quantum threats as they change. The proposed answer is a flexible way to handle an extensive use cases in the blockchain space

ble for multiple entities to participate while ensuring the confidentiality of their transactions.

- The quantum-hybrid blockchain is a novel tactic that integrates the benefits of quantum-secured and quantum-enhanced blockchains. This fusion results in the development of a system that exhibits robustness, security, and a high degree of resilience.

It is imperative to acknowledge that the investigation of quantum blockchain is now in its nascent stage, and the most effective configuration, if one exists, has yet to be ascertained. Furthermore, the practical ramifications of this technology remain incompletely actualized, requiring a significant duration before it can attain its maximum efficacy.

2.4   Proposed Algorithms for Quantum Blockchain

Several algorithms have been proposed in current research endeavors for the creation of a quantum blockchain. The subsequent examples serve to exemplify a range of essential algorithms. The provided text encompasses a range of numerical data.

- Quantum key distribution (QKD) is a technique that utilizes the moralities of quantum physics to launch a secure means of communication across many entities. Subsequently, the establishment of this safe channel can be employed to produce encryption keys, ensuring the security of blockchain transactions.
- Quantum-secured direct communication (QSDC) is a scheme that leverages the phenomenon of quantum entanglement to obviate the necessity of a specialized secure communication channel, hence facilitating direct communication among several entities. The utilization of this approach possesses the capability to enhance the confidentiality of transactions executed inside a quantum blockchain environment.
- The quantum hash function (QHF) is a method that utilizes the fundamental moralities of quantum physics to generate a hash function that is resilient against collisions and preimages. The implementation of this method possesses the capacity to augment the security of the blockchain by augmenting the amount of complexity linked to the alteration or manipulation of the data that is documented on the blockchain.
- The quantum random number generator (QRNG) exploits the foundational moralities of quantum mechanics to generate random numbers. Numerical values can subsequently be employed to produce robust cryptographic keys, hence enhancing the level of security offered by blockchain technology.
- Grover's algorithm is a quantum computational approach that exhibits the capability to speed up locating a particular blockchain transaction by diminishing the number of necessary searches to N, wherein N is the magnitude of the database.

It is imperative to acknowledge that most of these algorithms are now in the research stage and have yet to be applied in real settings. Furthermore, the most effective approach for implementing the quantum blockchain remains uncertain. Furthermore, the incorporation of quantum algorithms into blockchain systems is a significant barrier that necessitates considerable investments of time and resources for its advancement.

3   Quantum Cryptography

Quantum cryptography, alternatively mentioned to as quantum key distribution (QKD) [23], is a method employed for the protected transmission of cryptographic keys by the application of principles derived from quantum physics. The technique facilitates the establishment of a mutually agreed upon secret key between two entities, which can subsequently be utilized for the purpose of encrypting and decrypting messages exchanged between such parties. The robustness of the key's security is derived from the essential moralities of quantum physics, rendering it highly challenging for an unauthorized party to pilfer it without triggering detection mechanisms. Quantum cryptography is based on the essential moralities of quantum physics, particularly those related to the polarization and phase characteristics of light in quantum states. The quantum state of a photon contains the encryption key, which is then exchanged between two entities. Any endeavors to apprehend the photon will unavoidably modify its quantum state, thereby notifying the individuals engaged in a dialog that an act of surveillance is being conducted. Quantum cryptography exhibits superior performance compared to conventional cryptography across multiple dimensions. Therefore, the system demonstrates a significant degree of resilience against a wide range of attacks, including those executed by quantum workstations. Moreover, the computational ability of computers to explain mathematical problems is irrelevant to the current topic under discussion. Hence, it may be inferred that forthcoming developments in processing capacity would not have any discernible impact [24, 25].

However, it is imperative to recognize that quantum cryptography encounters specific obstacles, particularly in its implementation within real-world scenarios. The methodology exhibits sensitivity to extraneous disturbances and necessitates the utilization of reliable and robust equipment. Furthermore, it is worth noting that there is a finite distance beyond which a quantum key distribution (QKD) system is unable to reliably convey a secure cryptographic key. Quantum cryptography is a cryptographic methodology that employs the underlying moralities of quantum physics to guarantee the secure conversation with cryptographic keys. This methodology provides a heightened level of security in comparison with traditional procedures. The technology under consideration possesses the prospective to function as an essential element in the advancement of blockchain systems that are immune to quantum attacks, as well as other communication systems that prioritize security.

3.1   Quantum Cryptography vs. Traditional Cryptography

There are major distinctions between traditional cryptography and quantum cryptography, even though both seek to guarantee the data confidentiality through encryption and transmit that data. In the traditional form of cryptography, the process of encrypting and decrypting data takes place via the application of various mathematical methods. This subfield of cryptography is sometimes mentioned to as "classical" cryptography, particularly in specific circles. The success of these algorithms is dependent on the completion of complex mathematical operations, such as the division of a very big integer or the solution of discrete logarithms. Consider, for instance, the operation of dividing discrete logarithms by a significant number. This can serve as an illustration. Traditional cryptography has proven to be effective in the past, but it is vulnerable to the potential dangers posed by quantum computers.

When it comes to the processing power they bring to bear on these mathematical challenges, quantum computers handily beat out their more traditional counterparts. To carry out its operations, quantum cryptography makes use of the fundamental laws that govern quantum physics. Encryption of data is a part of this process, and it does so by utilizing the characteristics of quantum states, more specifically polarization and phase. Using the limitations imposed by quantum mechanics in the arena of quantum cryptography makes it impossible to steal a cryptographic key in a covert manner because any attempt to do so would always be discovered. This is because quantum mechanics imposes these limitations. Data encryption using quantum cryptography is regarded as among the safest methods available. The term "secure" is commonly used to refer to quantum cryptography, which implies that it is resistant to any type of decryption, including that which is performed by quantum computers [26, 27].

Since they are based on mathematical principles, traditional encryption techniques are vulnerable to being broken by quantum computers. Quantum cryptography, on the contrary side, is founded on the theories of quantum mechanics, which makes it fundamentally safe in every and every circumstance.

3.2 Research Gaps in Quantum Cryptography

The field of quantum cryptography is now experiencing significant growth and is marked by some unresolved matters that necessitate more investigation and solutions. There are several significant research challenges that exist within the domain of quantum cryptography, encompassing the following concerns:

- *Distance limitations:* The field of quantum key distribution (QKD) encounters a notable challenge in the shape of limitations imposed by distance. Current quantum key distribution (QKD) systems suffer from constrained transmission ranges principally caused by signal attenuation over long distances. The practical uses of quantum key distribution (QKD) are significantly constrained by the restricted transmission distance it can achieve. Hence, scholars are actively involved in the formulation of approaches to overcome this constraint and enhance the extent of transmission.
- *Noise:* The impact of noise on quantum key distribution systems can compromise their effectiveness by introducing faults into the transmitted key. The existence of various sources of noise in practical scenarios might provide a substantial barrier, impacting the quality and reliability of the signal. Academic scholars are currently directing their efforts toward developing mitigation measures for the adverse impacts of noise in quantum key distribution (QKD) systems.
- *Scalability:* Quantum key distribution (QKD) schemes presently exhibit restricted scalability and lack the capability to accommodate a significant number of users. Currently, there are ongoing endeavors to augment the scalability

and practicality of quantum key distribution (QKD) systems for their application in practical circumstances.

- *Real-world implementation:* Theoretical security has been demonstrated in the field of quantum cryptography; nevertheless, the practical implementation of this technology in real-world scenarios remains unclear. Currently, researchers are actively involved in the endeavor of formulating effective approaches for the implementation of quantum cryptography in real-world scenarios.
- *Security against side-channel attacks:* Quantum cryptography is widely acknowledged for its inherent security that is considered impervious to decryption. Nevertheless, the level of vulnerability of the system to side-channel attacks, which use implementation specifics or physical properties to harvest personal data, is still undetermined.
- *Quantum repeater:* The limited distance capacity of existing quantum key distribution (QKD) systems poses a substantial challenge for the transmission of quantum keys over large distances. The quantum repeater intentions to overcome the constraint of limited distance by employing a sequence of intermediary nodes to transmit the quantum signal. The primary function of these nodes is to facilitate the transmission an amplification of the signal. Nevertheless, the investigation of this notion is still in premature phases, and its practical application has yet not been comprehended.

The concerns encompass a range of unresolved obstacles in the domain of quantum cryptography, compelling researchers to actively seek resolutions in order to enhance the practicality and widespread implementation of this cryptographic technique.

3.3 *Advantages of Quantum Cryptography*

Quantum key distribution, also referred to as quantum cryptography or QKD, exhibits several distinct advantages when compared to traditional cryptographic techniques. Quantum cryptography has been extensively discussed in scholarly literature [28–30], whereby several noteworthy advantages of this field have been highlighted.

- *Unconditional security:* The term "unconditional security" pertains to a condition of complete certainty or safeguarding that is not contingent upon any specific criteria or prerequisites.
- *Quantum cryptography* is generally recognized for its inherent security, rendering it invulnerable to various forms of attacks, including those that may be generated by quantum computers. The technology in question demonstrates significant promise in enhancing the security of communication and transactions in the future, especially considering the anticipated advancement of quantum workstations that will outperform computational power over classical systems.

- *Tampering detection* is the term used to describe the procedure of identifying and detecting any unauthorized alterations or modifications that have been done to a system or its components.
- *Quantum cryptography* facilitates the identification of any illegal interception or alteration of the transmitted key, a feat unattainable by conventional cryptographic techniques. This characteristic renders it a very secure mode of communication.
- *Randomness:* Quantum cryptography utilizes principles derived from quantum physics to generate numbers that are truly random, thereby enabling the creation of cryptographic keys with enhanced security. In contrast to conventional cryptographic methods that depend on pseudorandom number generators, this characteristic enhances its security.
- *Quantum resistance:* The concept of "quantum resistance" pertains to the ability of a cryptography system to survive adversarial attempts launched by quantum computers. As quantum computing progresses, the processing competences of quantum computers pose a possible vulnerability to conventional cryptography schemes. In contrast, quantum cryptography possesses the facility to withstand the computational power of quantum computers, rendering it an appealing choice for ensuring enduring security.
- *No secret sharing:* The notion of "no secret sharing" pertains to the abstention from divulging or disseminating privileged data. In the realm of traditional cryptography, the distribution of keys to all pertinent entities is of extreme prominence. In the realm of quantum cryptography, the process of key establishment is limited to a bilateral exchange between two individuals or entities, hence obviating the need for involvement from a third party. This characteristic enhances the overall security of the system.
- *No need for pre-distribution:* In the realm of conventional cryptography, the utilization of keys necessitates their sharing and dissemination, hence imposing a substantial constraint. In contrast, quantum cryptography facilitates the concurrent production and distribution of cryptographic keys, resulting in enhanced operational efficiency.

Quantum cryptography has several notable advantages in comparison with conventional methodologies employed to address the same issue. Although, the tremendous benefits provided by quantum cryptography persist across various applications. Moreover, it is essential to recognize that this technology is still in the early stages of development, and numerous obstacles must be overcome before widespread application.

3.4 Disadvantages of Quantum Cryptography

Quantum cryptography: quantum key distribution (QKD) possesses the various benefits over the traditional cryptographic algorithms. However, it does have some limitations. Among the principal disadvantages of quantum cryptography are as follows [31, 32]:

- *Complexity:* Due to the need for specialized apparatus and knowledge, the implementation of quantum cryptography presents significant obstacles. In addition, the successful application of this technology necessitates a comprehensive understanding of the fundamental moralities of quantum mechanics, posing adoption challenges for certain businesses.
- *Cost:* The enactment of quantum cryptography requires the use and maintenance of specialized hardware and software, which may incur significant expenses. In addition, the implementation of quantum cryptography can incur substantial expenses, especially in comparison with conventional encryption methods.
- *Noise:* Quantum cryptography systems are susceptible to noise, which can lead to defects in the transmission of cryptographic keys. In certain circumstances, the presence of many sources of noise might provide a substantial hindrance.
- *Existing quantum key distribution (QKD)* systems have a limited transmission range due to signal intensity degradation over long distances. Consequently, the range capabilities of these systems are limited. Quantum key distribution (QKD) has extremely limited practical applications, and the investigation of quantum repeaters to increase its utility is in its earliest stages of development.
- *Scalability:* Quantum key distribution (QKD) systems in their current state exhibit limited scalability and user capacity. The circumstance can be problematic for large organizations that must ensure communication for many users.

Quantum cryptography is an emerging arena of technology that is presently inaccessible to the public on a wide scale. In addition, it is essential to recognize that the contemporary technology is still in its embryonic stage, indicating that the process of achieving widespread adoption could take a considerable quantity of time. Quantum cryptography has several significant disadvantages compared to conventional encryption methods. Quantum cryptography, despite its inherent limitations, possesses the property of absolute security and has the impending to provide a sophisticated level of security in certain application scenarios compared to conventional methods.

4   Consensus Algorithms

Consensus algorithms are employed in blockchain systems to guarantee unanimous agreement among all participants in the blockchain network. A wide range of algorithms have been developed to attain consensus, each with its own distinct set of strengths and weaknesses [33].

It is remarkable. The utilization and widespread implementation of the proof-of-work (PoW) consensus method in many blockchain systems exemplify this phenomenon. In this process, miners are obligated to solve intricate mathematical problems with the purpose of appending new blocks to the blockchain. Quantum computers possess the ability to do operations at a significantly accelerated pace compared to traditional computers. The increased processing speed possesses the capacity to expedite the resolution of mathematical issues in proof-of-work (PoW) protocols, surpassing the computational efficiency of conventional computers. The utilization of quantum computers possesses the capacity to significantly enhance the speed of block mining in comparison with traditional computers. Consequently, this acceleration may result in the consolidation of mining capabilities among a limited group of miners equipped with quantum-enabled technology. Consequently, the concentration may give rise to a singular point of vulnerability, thereby posing a potential security hazard [34].

On the other hand, other consensus algorithms, such as Byzantine fault tolerance (BFT) and proof-of-stake (PoS), exhibit a reduced vulnerability to the potential impacts of quantum computing. This phenomenon can be attributed to their dependence on procedures that extend beyond computational capacity, such as token ownership or consensus achieved by a defined number of validators [35].

The recognition of the probable impact of quantum computing on the efficacy of specific consensus algorithms is of utmost significance. Nevertheless, it is imperative to acknowledge that quantum computing remains a dynamic discipline, wherein various potential solutions are now being investigated and refined through ongoing research and development efforts. Furthermore, it is anticipated that the progression of quantum computing to a level where it can proficiently target blockchain systems will necessitate a substantial duration.

4.1   Proof-of-Work (PoW)

The consensus process recognized as proof-of-work (PoW) is extensively employed across several blockchain platforms. The major purpose of this system is to verify transactions and maintain the network's integrity. To address an intricate mathematical problem, individuals, referred to as "miners," partake in a competitive manner within a blockchain framework that runs based on the proof-of-work (PoW) principle. The initial miner who successfully fulfills this task is authorized to append a novel block of transactions to the blockchain. To provide a consistent development rate for the chain of blocks, the level of difficulty associated with this process is modified accordingly [36].

The potential impact of quantum computing on blockchain systems that depend on the proof-of-work consensus mechanism is substantial. The increased rate at which quantum computers may execute specific operations has the potential to drastically reduce the time required to solve proof-of-work (PoW) mathematical difficulties, in comparison with conventional computers. The capabilities may potentially allow quantum computers to speed up mining blocks in comparison with classical computers. Consequently, there is a possibility that this could result in a consolidation of mining authority among a limited group of miners equipped with quantum technology. Enabling the functionality to control the network and establish a singular point of failure may potentially introduce a security vulnerability. Currently, researchers are actively engaged in the advancement of proof-of-work (PoW) algorithms that are resistant to quantum computing. The objective is to enhance the security of blockchain networks by safeguarding them against possible dangers posed by quantum computers. The methods under consideration would rely on computationally challenging issues that are highly intricate for quantum computers, such as the learning with errors (LWE) problem [37].

The utilization of quantum proof-of-work (PoW) as a consensus process in blockchain networks serves to enhance their security by using principles derived from quantum computation. Figure 1 is an exemplar pseudocode example that pertains to a quantum proof-of-work (PoW) algorithm [38].

It is imperative to recognize that the potential influence of quantum computing on proof-of-work (PoW)-based blockchain systems is a significant subject matter. However, it is imperative to acknowledge that quantum computing is a nascent discipline, and a multitude of suggested approaches are now undergoing investigation in the research stage. Furthermore, the advancement of quantum computing to a degree where it can proficiently engage with proof-of-work (PoW)-based blockchain systems may need a significant duration.

4.2 Quantum Smart Contracts

Smart contracts utilize computer code to encapsulate the exact details of a transaction conducted between a buyer and a seller. The utilization of blockchain technology enables the storage and replication of contracts across the network. The network will subsequently authenticate the fulfillment of the tasks. The facilitation of contract execution automation can be achieved by the utilization of smart contracts, which possess the ability to carry out diverse supplementary tasks such as supply chain management, financial transactions, and voting systems. The utilization of artificial intelligence (AI) has promised to enhance the operational efficiency of enterprises and optimize supply chain operations through the elimination of intermediaries. The field of quantum computing holds promise for significantly enhancing the velocity and efficacy of specific computational endeavors, particularly those

```
# Quantum Proof-of-Work Algorithm

# Function to mine a new block
def mine():
    # Generate a random challenge
    challenge = generateChallenge()
    # Attempt to solve the challenge using quantum computing
    solution = solveChallenge(challenge)
    # Check if the solution is valid
    if (validateSolution(solution)):
        # Create a new block
        newBlock = createBlock(challenge, solution)
        # Add the new block to the blockchain
        addBlock(newBlock)
        # Broadcast the new block to the network
        broadcastBlock(newBlock)
    else:
        # Repeat the mining process
        mine()

# Function to generate a random challenge
def generateChallenge():
    # Use a quantum-resistant random number generator
    challenge = generateQuantumRandomNumber()
    return challenge

# Function to solve the challenge using quantum computing
def solveChallenge(challenge):
    # Use a quantum algorithm to solve the challenge
    solution = performQuantumComputation(challenge)
    return solution

# Function to validate the solution
def validateSolution(solution):
    # Check if the solution meets the required conditions
    if (solution is valid):
        return True
    else:
        return False

# Function to create a new block
def createBlock(challenge, solution):
    # Create a new block with the challenge and solution
    newBlock = {
        "challenge": challenge,
        "solution": solution,
        "timestamp": getCurrentTimestamp()
    }
    return newBlock

# Function to add the new block to the blockchain
def addBlock(newBlock):
    # Add the new block to the blockchain
    blockchain.append(newBlock)

# Function to broadcast the new block to the network
def broadcastBlock(newBlock):
    # Send the new block to all nodes in the network
    sendBlockToNodes(newBlock)
```

Fig. 1 Pseudocode snippet for quantum proof-of-work consensus algorithm [38]

pertaining to cryptography and blockchain technology. The utilization of quantum computing holds promises in enhancing smart contracts by enabling the execution of contractual agreements that exhibit heightened complexity and enhanced security. Quantum smart contracts (QSCs) represent an innovative notion that continues to be the focus of continuing scholarly investigation. Quantum-secured communications (QSCs) possess several prospective real-time uses, encompassing the following:

- Quality control systems (QCS) are of utmost importance in the field of supply chain administration, as they are responsible for safeguarding the genuineness and reliability of items during their journey across the supply chain. This measure possesses the capacity to diminish the probability of fraudulent actions and the manufacture of counterfeit products inside the supply chain.
- Quantum-secured communications (QSCs) possess the ability to facilitate secure and transparent transactions within the domain of financial services, encompassing areas such as banking, insurance, and investment. The application of suitable strategies can significantly decrease the occurrence of deceitful behaviors, therefore safeguarding the confidentiality and integrity of monetary transactions.
- Quality and security credentials (QSCs) possess the capacity to be employed within the healthcare sector for the purpose of safeguarding patient confidentiality, facilitating the secure exchange of data, and enabling secure transactions among healthcare providers.
- The integration of quantum-secured computation (QSC) into voting systems offers a means to guarantee the secure registration and aggregation of ballots while also facilitating transparency and verifiability at every stage of the voting procedure.
- Quantum-safe communication (QSC) has the potential to enhance the security of Internet of Things (IoT) devices by facilitating safe communication and data sharing across these devices.

The utilization of quantum-resistant algorithms has the potential to enhance the security of smart contract execution, therefore mitigating the vulnerability to unauthorized alterations. Moreover, the utilization of quantum computation possesses the capability to enhance the scalability of smart contract platforms by facilitating accelerated transaction processing. However, it is imperative to acknowledge that the realm of quantum computation is now in its early stages of advancement, and its potential impact on smart contracts has not been comprehensively comprehended [39]. It is imperative to acknowledge that the predominant approach to the development of smart contracts predominantly depends on conventional computing methods. The domain of quantum smart contracts, however, is now seeing ongoing research and development efforts. Figure 2 illustrates the pseudocode of the quantum smart contract, as presented in reference [40].

In summary, the quantum-secured blockchain (QSB) signifies a notable progression in the field of blockchain technology. The primary aim of this project is to provide enhanced levels of security and anonymity by integrating blockchain technology and quantum cryptography. The integration of quantum computing

```
# Function to initialize the contract
def init():
    # Set the initial state of the contract
    state = "initialized"

# Function to execute the contract
def execute(inputs):
    # Verify that the inputs are valid
    if (validateInputs(inputs)):
        # Perform the operations defined in the contract
        result = performOperations(inputs)
        # Use quantum-resistant algorithms to encrypt the result
        result = encryptResult(result)
        # Update the state of the contract
        state = "executed"
        # Return the encrypted result
        return result
    else:
        # Return error message
        return "Invalid inputs"

# Function to validate the inputs
def validateInputs(inputs):
    # Check if inputs meet the required conditions
    if (inputs are valid):
        return True
    else:
        return False

# Function to perform the operations defined in the contract
def performOperations(inputs):
    # Perform calculations using quantum computing
    result = performQuantumCalculations(inputs)
    return result

# Function to encrypt the result
def encryptResult(result):
    # Use quantum-resistant algorithms to encrypt the result
    encryptedResult = encryptUsingQuantumAlgorithms(result)
    return encryptedResult
```

Fig. 2 Pseudocode snippet for quantum smart contract [40]

capabilities into the field of blockchain technology has presented new opportunities for the advancement of more resilient and fortified blockchain solutions. The primary objective of this study was to examine the possible utilization of quantum-secured bit (QSB) technology within the domains of smart contracts, consensus mechanisms, and quantum cryptography. The benefits of QSB in comparison with traditional blockchain systems were emphasized, while the challenges that necessitate additional investigation were addressed. The potential of QSB appears great, given the continuous improvements in quantum computing and blockchain technology. The objective of this study is to offer relevant viewpoints to scholars and practitioners in this field.

References

1. Li, C., Xu, Y., Tang, J., & Liu, W. (2019). Quantum blockchain: A decentralized, encrypted and distributed database based on quantum mechanics. *Journal of Quantum Computing, 1*(2), 49.
2. Li, M., Wang, L., & Zhang, Y. (2021). A framework for rocket and satellite launch information management systems based on blockchain technology. *Enterprise Information Systems, 15*(8), 1092–1106.
3. Feng, M., & Xu, H. (2019, June). MSNET-blockchain: A new framework for securing mobile satellite communication network. In *2019 16th annual IEEE international conference on sensing, communication, and networking (SECON)* (pp. 1–9). IEEE.
4. Allende, M., León, D. L., Cerón, S., Leal, A., Pareja, A., Da Silva, M. & Venegas-Andraca, S. E. (2021). Quantum-resistance in blockchain networks. *arXiv preprint arXiv:2106.06640*.
5. Shor, P. W. (1997). Polynomial-time algorithms for prime factorization and discrete logarithms on a quantum computer. *SIAM Journal on Computing, 26*, 1484–1509.
6. Mosca, M. (2018). Cybersecurity in an era with quantum computers: Will we be ready? *IEEE Security and Privacy, 16*, 38–41.
7. Fedorov, A. K., Kiktenko, E. O., & Lvovsky, A. I. (2018). Quantum computers put blockchain security at risk. *Nature, 563*, 465–467.
8. Stewart, I., Ilie, D., Zamyatin, A., Werner, S., Torshizi, M., & Knottenbelt, W. (2018). Committing to quantum resistance: A slow defence for Bitcoin against a fast quantum computing attack. *Royal Society Open Science, 5*, 180410.
9. Gao, Y., Chen, X., Chen, Y., Sun, Y., Niu, X., & Yang, Y. (2018). A secure cryptocurrency scheme based on post-quantum blockchain. *IEEE Access, 6*, 27205–27213.
10. Li, C., Chen, X., Chen, Y., Hou, Y., & Li, J. (2019). A new lattice-based signature scheme in post-quantum blockchain network. *IEEE Access, 7*, 2026–2033.
11. Kiktenko, E. O., Pozhar, N. O., Anufriev, M. N., Trushechkin, A. S., Yunusov, R. R., Kurochkin, Y. V., Lvovsky, A. I., & Fedorov, A. K. (2018). Quantum-secured blockchain. *Quantum Science and Technology, 3*, 035004.
12. Aggarwal, D., Brennen, G., Lee, T., Santha, M., & Tomamichel, M. (2018). Quantum attacks on bitcoin, and how to protect against them. *Ledger, 3*.
13. Sun, X., Wang, Q., Kulicki, P., & Sopek, M. (2019). A simple voting protocol on quantum blockchain. *International Journal of Theoretical Physics, 58*, 275–281.
14. Rodenburg, B., & Pappas, S. P. (2017). *Blockchain and quantum computing*. The MITRE Corporation.
15. Fedorov, A. K., Kiktenko, E. O., & Lvovsky, A. I. (2018). Quantum computers put blockchain security at risk.
16. Fernandez-Carames, T. M., & Fraga-Lamas, P. (2020). Towards post-quantum blockchain: A review on blockchain cryptography resistant to quantum computing attacks. *IEEE Access, 8*, 21091–21116.
17. Banerjee, A. (2021). Blockchain vs. quantum computing: Is quantum computing the biggest threat to crypto? Blockchain-council.org, 08-Dec-2021. [Online]. Available: https://www.blockchain-council.org/blockchain/blockchain-vs-quantum-computing-is-quantum-computing-the-biggest-threat-to-crypto/. Accessed 17 Nov 2022.
18. Kiktenko, E. O., Pozhar, N. O., Anufriev, M. N., Trushechkin, A. S., Yunusov, R. R., Kurochkin, Y. V., Lvovsky, A. I., & Fedorov, A. K. (2018). Quantum-secured blockchain. *Quantum Science and Technology, 3*(3), 035004.

19. Quantum blockchains - quantum cryptography for blockchains. Quantum Blockchains. [Online]. Available: https://www.quantumblockchains.io/. Accessed 11 Oct 2022.
20. Wang, W., Yu, Y., & Du, L. (2022). Quantum blockchain based on asymmetric quantum encryption and a stake vote consensus algorithm. *Scientific Reports, 12*(1), 1–12.
21. Sun, X., Wang, Q., Kulicki, P., & Zhao, X. (2018). Quantum-enhanced logic-based blockchain i: Quantum honest-success byzantine agreement and qulogicoin. *arXiv preprint arXiv:1805.06768*.
22. Swan, M. (2020). Quantum blockchain. In *Quantum computing* (pp. 113–134). World Scientific.
23. Renner, R. (2008). Security of quantum key distribution. *International Journal of Quantum Information, 6*(01), 1–127.
24. Gisin, N., Ribordy, G., Tittel, W., & Zbinden, H. (2002). Quantum cryptography. *Reviews of Modern Physics, 74*(1), 145.
25. Quantum cryptography, explained. QuantumXC, 28-Nov-2018. [Online]. Available: https://quantumxc.com/blog/quantum-cryptography-explained/. Accessed 2 Dec 2023.
26. Moizuddin, M., Winston, J., & Qayyum, M. (2017, March). A comprehensive survey: Quantum cryptography. In *2017 2nd international conference on anti-cyber crimes (ICACC)* (pp. 98–102). IEEE.
27. Pirandola, S., Andersen, U. L., Banchi, L., Berta, M., Bunandar, D., Colbeck, R., Englund, D., et al. (2020). Advances in quantum cryptography. *Advances in Optics and Photonics, 12*(4), 1012–1236.
28. Jiang, L., Taylor, J. M., Nemoto, K., Munro, W. J., Van Meter, R., & Lukin, M. D. (2009). Quantum repeater with encoding. *Physical Review A, 79*(3), 032325.
29. What benefits does quantum cryptography provide in this era?. Educative: Interactive Courses for Software Developers. [Online]. Available: https://www.educative.io/answers/what-benefits-does-quantum-cryptography-provide-in-this-era. Accessed 4 Oct 2022.
30. Zhou, T., Shen, J., Li, X., Wang, C., & Shen, J. (2018). Quantum cryptography for the future internet and the security analysis. *Security and Communication Networks, 2018*, 1–7.
31. Bennett, C. H., & Brassard, G. (1985). An update on quantum cryptography. In *Workshop on the theory and application of cryptographic techniques* (pp. 475–480). Springer.
32. Mitra, S., Jana, B., Bhattacharya, S., Pal, P., & Poray, J. (2017, November). Quantum cryptography: Overview, security issues and future challenges. In *2017 4th international conference on opto-electronics and applied optics (optronix)* (pp. 1–7). IEEE.
33. Porat, A., Pratap, A., Shah, P., & Adkar, V. (2017). Blockchain consensus: An analysis of proof-of-work and its applications.
34. Kumar, A., & Jain, S. (2021). Proof of game (PoG): A proof of work (PoW)'s extended consensus algorithm for healthcare application. In *International conference on innovative computing and communications* (pp. 23–36). Springer.
35. Jain, S., & Kumar, A. (2022). A security analysis of lightweight consensus algorithm for wearable kidney. *International Journal of Grid and Utility Computing, 13*(5), 505–525.
36. Bard, D. A., Kearney, J. J., & Perez-Delgado, C. A. (2022). Quantum advantage on proof of work. *Array, 15*, 100225.
37. Aggarwal, D., Brennen, G. K., Lee, T., Santha, M., & Tomamichel, M. (2017). Quantum attacks on Bitcoin, and how to protect against them. *arXiv preprint arXiv:1710.10377*.
38. Bashir, I. (2017). *Mastering blockchain*. Packt Publishing Ltd..
39. Khan, S. N., Loukil, F., Ghedira-Guegan, C., Benkhelifa, E., & Bani-Hani, A. (2021). Blockchain smart contracts: Applications, challenges, and future trends. *Peer-to-Peer Networking and Applications, 14*(5), 2901–2925.
40. Paleka, I. (2022). Create and test smart contracts using Python. Algorand.org. [Online]. Available: https://developer.algorand.org/tutorials/create-and-test-smart-contracts-using-python/. Accessed 15 Nov 2022.

The Intersection of Blockchain Technology and the Quantum Era for Sustainable Medical Services

Dinesh Kumar Atal ⓘD, Vishal Tiwari ⓘD, Anjali ⓘD, and Rajiv Kumar Berwer ⓘD

Abstract Medical data has become crucial to the healthcare sector's growth and development due to its complexity and expense of preserving security and confidentiality and providing services to the entire healthcare ecosystem. Therefore, an open-channel platform on the Internet enables patients to manage, share, and keep track of their electronic health records (EHRs) with friends, family members, and healthcare professionals. After that, the growth of medical data comes with the concern of handling it securely. So, for this, blockchain technology significantly simplifies the process of conducting transactions, as it can store data in vast amounts in a distributed manner which may be retrieved wherever and whenever needed, thereby improving the efficiency of today's healthcare system. Medical records processing using quantum and blockchain technology heavily influenced the industrial healthcare market. These technologies can benefit healthcare and significantly affect its transparency, information security management, business opportunities, and processing efficiency. Quantum blockchain technology is another tool that can enhance the speed of diagnosis and treatment because any delays in treatment or emergency care can compromise the security and confidentiality of medical records. This way, the medical records can be analyzed and communicated with privacy, security, availability, and authenticity. Quantum blockchain and computing have the capacity to acquire thermal imaging speedily with which patients can be located and monitored simultaneously, thereby optimizing diagnosis expenses and insurance fraud. This review chapter discusses how smart healthcare systems benefit from quantum and blockchain.

D. K. Atal (✉) · V. Tiwari · Anjali
Department of Biomedical Engineering, Deenbandhu Chhotu Ram University of Science and Technology, Sonipat, Haryana, India

R. K. Berwer
Department of Computer Science and Engineering, Deenbandhu Chhotu Ram University of Science and Technology, Sonipat, Haryana, India

S. Pulipeti et al. (eds.), *Quantum and Blockchain-based Next Generation Sustainable Computing*, Contributions to Environmental Sciences & Innovative Business Technology, https://doi.org/10.1007/978-3-031-58068-0_2

Keywords Healthcare industry · Electronic health records · Blockchain technology · Quantum computing · Quantum and blockchain technology

1 Introduction

The growth of industrial healthcare comes with many digitally recorded medical databases, which require secure processing. The adoption of blockchain and IoT (Internet of Things) significantly impacts the industrial healthcare sector owing to the industry's design comprising of related apparatus and IT applications. These technologies will enhance information security, transparency, regulatory regulation, and processing efficiency while opening up new commercial prospects [1]. Therefore, healthcare information could be assessed, evaluated, and shared and maintain patient privacy. Blockchain technology in modern medicine has many benefits, such as data management, accounting, and finance. Additionally, there are specific issues and dangers related to blockchain-based healthcare projects and suggestions for mitigating or eliminating them. The health industry is one of the areas where the blockchain can substantially influence on both the economy and society [2, 3]. The most significant benefit using it does not require the participation of any centralized organization. In the medical domain, blockchain technology has been used to share clinical data, account management, drug supply chain management, drug development, and clinical trials. The interchange of clinical information between the payers, patients, and partners and how we obtain and maintain clinical details can all be entirely transformed by blockchain technology [2].

The rapidly developing technology, blockchain, is utilized in various security applications. The Stakeholders employ a variety of implementations that use blockchain technology that kept as a sequence of blocks, may serve as a public record for all committed transactions. Some essential features of blockchain technology contain decentralization, availability, immutability, transparency, and persistence [4]. Blockchain technology can potentially improve the clinical data-sharing aspect of the IoT (Internet of Things). Blockchain provides a secure means of distributing critical patient information collected by IoT devices [1, 5]. The BPIIoT (Blockchain Platform for Industrial Internet of Things) enables the legacy shop floor device integration with an appropriate cloud environment; it would show a tremendous increase in performance over the current CBM (cloud-based manufacturing) platform, removing financial barrier that has stifled the development of a company ecosystem. Using this ecosystem, the manufacturing and healthcare industries, including supply chains, logistics, healthcare, agriculture, and the energy sector, may be able to facilitate transactions between users [1, 6, 7].

Several healthcare institutions utilize the IoHT (Internet of Healthcare Things) for maintaining assets, monitoring infants, and tracking inventory [1, 8]. The use cases are two types: assistance operations and clinical services. Through RPM (remote patient monitoring), IoHT facilitates patient-centric activities in healthcare. Internet of Healthcare Things strictly checks all vital signs and other crucial parameters for the investigation, like weight and blood sugar level changes [1]. The increased utilization of mobile medical assets made possible by IoHT enhances support operations and lowers overall operating expenses. The staff can access real-time information on the location and usage patterns of ventilators, digital X-ray equipment, and other resources while saving money due to equipment sensors and data collection capabilities. This allows for more effective equipment assignment and faster location whenever and wherever needed, thereby saving valuable time for care workers. Using IoHT sensory inputs, expensive machinery such as MRI (magnetic resonance imaging) equipment can also be displayed to technicians in real time using IoHT sensory inputs. To develop digital twins of technology, medical decision-makers are also exploring integrating augmented reality technology and IoHT [1, 9]. The workers and clinicians are provided with practical, hands-on training opportunities and highly visual and interactive augmented reality interfaces that can digitally reproduce complicated medical equipment and devices. The protection, security, and dependability of the servers that link IoT devices and share vital medical records are issues to be aware of when adopting IoHT [1, 10]. A blockchain is a logical option that can completely secure the entire process through encryption and decentralization [1, 11]. Blockchain is gaining popularity due to its ability to create and disseminate permanent, irrevocable transactions securely. Blockchain builds blocks of transactions and stores them in a continuous sequence of occurrences shared securely with different participants. The documents are nearly impossible to alter since the blocks are secured using cutting-edge cryptographic technology [1, 12]. This technology keeps the possibility to streamline and progress security and precision for laborious, wasteful methods [1, 13]. Examples include the following:

- Quick medical insurance enrolment
- More streamlined claims adjudication
- Increased B2B movement throughout the healthcare value chain

The quicker and more effective certification of employees is one of the additional blockchain prospects. The healthcare sector is a good fit for IoHT and blockchain technologies due to its patient-centered mechanisms [1, 14]. The two technologies working together permit for the secure, unchangeable conversation of medical data [1]. Blockchain technology can benefit the medical field and battle the COVID-19 pandemic. Blockchain technology and its processes will indeed be employed in upcoming smart healthcare schemes for collecting relevant information from secure data storage, sensors, and automatic patient monitoring, according to the significance of the blockchain. Blockchain technology significantly streamlines operations since it can preserve vast amounts of information in a distributed and more secure manner. It provides a more excellent range of accessibility whenever and wherever necessary [15]. Quantum computing, on the other hand, provides access to excellent services. The quantum computing has several advantages including the capacity to employ thermal imaging based on quantum computing and the ability to detect and monitor patients quickly. These advantages can be completely realized when combined with quantum blockchain [15]. Another technique that can protect the reliability, confidentiality, and accessibility of data records is quantum blockchain. If quantum computing and blockchain technologies are combined, medical records processing will be faster and more private [15]. Therefore, the use of quantum blockchain technology would benefit medical services to a great extent. The pace of diagnosis and treatment can be accelerated with quantum blockchain technology. It also helps preserve data records' availability and authenticity. The advantages of quantum computing, like the capacity to obtain thermal imaging and the speed with which patients can be discovered and tracked, may all be utilized to their fullest extent with the help of quantum blockchain. Additionally, it maximizes insurance fraud and diagnosis costs.

2   Smart Healthcare Systems

Information technology (IT) is crucial in the healthcare industry because population health management technologies such as remote patient monitoring (RPM) and electronic health records (EHRs) have highly improved this industry. The medical records engendered from these references are extensive and time-consuming, resulting in issues with data quality, thereby making the data analysis, prediction, and diagnosis more complicated and the threat of data protection due to increased cybercrime [16]. Because they provide essential data consistent with patients' perspectives, medical and healthcare records have demonstrated their value to patients. The healthcare data repository may become a target and a moment of loss for system attackers. It increases the risk of vulnerabilities, leading to denial-of-service (DoS) and ransomware attacks [16]. The sharing of the patients's medical records and information among the healthcare providers via electronic health records may increase diagnostic performance. In particular, difficulties with patients' informed consent and the scientific credibility of findings such as data missing and dredging, endpoint flipping, and selective publication in clinical trials could be resolved by blockchain technology. A patient's brief medical history is included in an EHR along with statistics, projections, and any information/data related to their ailments and clinical progress throughout the therapy [17]. Users could access and preserve their health data using a blockchain system for EHRs that simultaneously ensures confidentiality and privacy [17].

EHRs allow patients to manage and share their health records with their family, medical professionals, and friends, over an open channel and the Internet. Despite the fact that cloud-based EHRs address difficulties, they are still susceptible to a variety of destructive attacks, server non-repudiation, and trust management. Hence, blockchain-based EHR solutions instill in consumers a sense of privacy, security, and trust [18, 19]. Implementing a blockchain-based EHR system that promotes interoperability and trust among all parties is possible. It has a distributed, time-stamped, chronological, immutable, and auditable log to keep clinical information. Numerous industries, including finance [20], education [21], edge computing services [22], tourism [23], automation [19], etc., have used blockchain technology.

The quantity of data and records has doubled in the healthcare organization 4.0, which has a $50 billion market [24].

Additionally, the percentage of verified electronic health record systems has doubled, from 42% to 87% [25]. Because medical professionals, hospitals, and other service providers update their records frequently, the considerable volume of data created from several sources could be more organized. It leads to issues related to management, such as drug tracking, bill management, and claim settlements. The Health Insurance Portability and Accountability Act (HIPPA) stipulates that healthcare industry stakeholders should ensure that approved data is uploaded by patients who have certified electronic health records (EHRs) from physicians. Privacy and data ownership concerns about patients' sensitive data, its storage, scalability of mined transactions, capacities, the cost of maintaining healthcare blockchain, and quantum and collusion attacks are the difficulties of using blockchain technology. Lattice-based cryptography offers the defense against quantum and collusion operations and the perfect answer to the issues of user privacy and data ownership. In the random oracle concept, lattices create post-quantum blockchain networks (P-QBN) observed as safe [26]. Therefore, lattice signature creation and verification activities authorize certified EHRs. According to HIPPA regulations, this guarantees complete record secrecy. Distributed artificial intelligence can enhance blockchain operations by addressing the scaling challenges of transactions, storage, and fee issues with digital health records stored digitally among distributed nodes [27]. Former conventional machine learning (ML) and statistical methods [28] were also employed likewise, whereas they examined plain data to discover relevant attributes and build patterns of initial intrigue. Identifying the criteria of interest is a much time-consuming skill [29]. Deep learning (DL) techniques, on the other hand, learn about feature selection with no human interference, allowing the finding of embedded dynamic relations in between the information [18]. The benefits of blockchain-based EHRs are numerous; some of them are records stored in a distributed manner (open and straightforward to verify across a wide range of unaffiliated provider organizations). Also, there is no centralized party for the attacker to crack or alter the data; information is continually updated and made available, and information from various means is combined into a single suitable information storage (Grey Healthcare Group, 2017) [17].

3   Blockchain Technology

Today's smart healthcare data management systems face serious hurdles regarding security, privacy, trust, flexibility, provenance, audit, traceability, transparency, and immutability. Additionally, many of the current smart medical record managing methods are centralized, which increases the danger of a single fault point. The blockchain, an emerging and disruptive decentralized technology, has the potential to radically revolutionize, reconfigure, and alter data management in medical domain [30]. Blockchain supports transaction audits in this way, going beyond transparency [17]. Blockchain is an innovation that encourages value sharing. It has recently been used in several fields, with finance in healthcare being the most significant. Blockchain is a distributed database of cryptographically linked blocks aggregating transactions sequentially. Each block is added to the ledger over peer-to-peer (P2P) networks, linked to the previous block using hashing technique. These records are built on a decentralized technology, eliminating the need for transactions and assets to interact with a centralized intermediary. Digital purchases, such as individual medical information, economic transactions, and personal records, can be processed, encrypted, confirmed, and stored more efficiently by emerging blockchain technology [31, 32].

Blockchain enables the decentralization and deregulation of markets of all transactions between participants daily. Blockchain is an application layer that advances the OSI layer by introducing a layer that allows instant digital currency and financial transactions. It maintains a record of all the completed transactions [31]. The chain will continue to expand after another transaction is completed. No third-party software is required for transactions. From a business perspective, the blockchain is a system for exchanging assets, value, and transactions between individuals. It is a method for validating and verifying the transactions and taking the place of trusted parties' entities [4]. Nowadays, every medical and healthcare system has high dependability on blockchain technology. Blockchain technology, which is decentralized and distributed, offers security services for the healthcare industry. The existing healthcare system's centralized design could be more secure across many medical services, resulting in delayed access to the concerned information and significant danger of information leakage. The medical records could then be archived without the client's concerns. The critical problem in the present healthcare maintenance system is gaining secure network access to information. Blockchain technology is proving to be an up-and-coming technology that is the most effective way to access the concerned data/information. When using blockchain technology, data is kept in a ledger feature that may be used to monitor how patients access their medical records [4].

By enabling unparalleled data efficiency and ensuring trust, blockchain helps streamline the various healthcare data management processes [5, 30]. It provides a vast range of built-in conspicuous capabilities, including decentralized storage, adaptability of data access, interconnection, authentication, security, immutability, and transparency. It also enables wider adoption of blockchain for managing healthcare information records [30, 33, 34]. Blockchain employs the idea of "smart contracts," which establish specific rules accepted by all the partners in the blockchain environment and eliminate the need for an intermediary [30, 35, 36]. It reduces unnecessary administrative costs. Consensus mechanisms, public-key cryptography, and peer-to-peer networks comprise the main three ideas of blockchain [27]. Public, private, and consortium blockchains are the categories into which blockchains are separated based on managing permission [30, 37]. Blockchain technology is a distributed database that handles the data within an IoT application in a chain of blocks [38, 39]. It is particularly good at securely processing relevant information for serverless edge computing. However, block data handling decreases processing speed [38, 40]. The procedure must operate concurrently over several distributed architecture nodes to verify the proof-of-work (PoW). Such systems might be integrated into a function as a service (FaaS) platform using microservices, which could be set up on a serverless pipeline. Quantum computers can provide large-scale resource management computations to address this issue [38, 41].

With numerous applications in areas like sharing patients' medical data, precision medicine, clinical trials, drug counterfeiting, longitudinal healthcare records, user-oriented medical research, public healthcare management, online patient access, and automated health claims adjudication, blockchain is essential to the healthcare industry [17]. Therefore, without needing third-party tools, blockchain technology enables more robust and more secure transactions. There is no further need for trust between the parties when each party to a transaction may believe in the correctness and consistency of the records, a concept known as "trustlessness" in the context of blockchain [1, 40]. Blockchain offers a broadly distributed, peer-validated, and immutable digital ledger and does not need money to operate correctly. Most enterprise-level blockchain applications do not call for specific money, coin, or token [1, 42].

3.1 *Types of Blockchain*

- *Private and public.* Who has access information to the blockchain database? Large audiences, the general public, can upload data to the ledger via public blockchains. A public blockchain network like Bitcoin illustrates this; there are no restrictions on who can exchange Bitcoin. Bitcoin can be purchased, sold, or sent to anyone. An illustration of a private solution would be a blockchain system that tracks how donations to nonprofit organizations are used. Only appointed authorities from the nonprofit group can disclose how funds are distributed and used in such a way [1, 43].
- *Permissioned and permissionless.* Permissionless systems are available, and the public needs more permission or role-based access. These platforms need the natural functionality to handle identities, set permissions based on those identities, and enforce those identities. This implies that you must design and implement a system to track and manage identification and draw permissions against that identity if you decide to do so. This does not suggest that you cannot develop a permissioned solution on a permissionless platform. Considering whether all participants should be treated equally or some should have access to features or rights that others do not is a beautiful approach to narrowing down the type of blockchain required when designing a solution. The solution to this question will assist in informing whether to employ permissioned or permissionless blockchain technology [1, 43].

3.2   Advantages and Limitations of Blockchain Technology

3.2.1   Advantages of Blockchain

A business network's organizations can share infrastructure owing to blockchain technology. Since we lose the fact if the data is altered, corrupted, or hacked, blockchain is more secure than a typical database. Blockchain offers robust security and a wide range of permissions to validate and regulate who has a permit to access information and under which conditions. Tracing the sources of all supply chain pieces also enhances quality assurance services by reducing costs and controlling any defective parts [1, 44, 45]. Blockchain is fault-tolerant and redundant; if any system loses track of the database, it will stay elsewhere on the network. Consider sending a group message to understand fault tolerance better. Since every member in the messaging party has a replica of the party conversation, they would have to do so on everyone else's phone if any member wanted to delete something from the conversation. When numerous people are involved, fault tolerance is beneficial. The tokenization, which opens up new entrepreneurial opportunities and produces tradeable tokens with real-world value, is another significant benefit. The term tokenization is the digitizing representation of fractional asset ownership, such as owning one car in one city or one hundred vehicles in one hundred. Blockchain uses a smart contract to ensure that business operations are automated and consistent across several enterprises. Finally, eliminating intermediaries by blockchain lowers costs and improves the effectiveness of company operations. It enables firms to function more quickly and respond to changes in the commercial environment considerably more rapidly than they might otherwise [1, 44, 45].

Patients can transfer their medical records to anyone without worrying about data tampering or corruption because the blockchain is immutable and traceable [4]. The same is true for a blockchain-generated and added medical record, which will be completely secure. The patient can influence how the institutions utilize and disseminate their medical data. Any party seeking access to the patient's medical records might use the blockchain to verify their eligibility [4]. A reward system can also motivate the patient to behave well. For instance, people may receive rewards for adhering to a care plan or maintaining good health. Additionally, they can receive tokens in exchange for providing their data for studies and clinical trials [4]. Due to the nature of the products they carry, pharmaceutical businesses must have a very secure supply chain. Drugs for the pharmaceutical industry are often stolen from the supply chain and sold illegally to various customers. Additionally, these businesses lose almost $200 billion a year due to the sale of counterfeit pharmaceuticals. These businesses will benefit from a transparent blockchain by closely tracking drugs back to their point of origin, which will help remove fake medicine [4].

3.2.2 Drawbacks of Blockchain

Like any other technology, blockchain has costs associated with its advantages. Its disadvantages need to be fully considered to decide whether blockchain would be the best option for an overall solution architecture. Best practices suggested patterns, and appropriate use cases are continually being established [1, 44, 45]. The blockchain's scalability in comparison with traditional technology is another significant limitation. Blockchain has a lower transaction volume processing capacity than competing systems like Visa. For blockchain to be performance competitive, the ability needs to be increased by several orders of magnitude. Additionally, obtaining a thorough "God Mode" view of the answer and its information can be difficult or impossible. Many available media and toolkits are yet in the preproduction stage and might not be ready to develop complex applications [1, 44, 45]. Energy consumption is a vital factor in the developing stage of blockchain technology. Proof-of-work (PoW), the initial agreement algorithm utilized to mine Bitcoin, uses 7.67 gigawatts of electricity annually. This power is comparable to the power use of nations like Austria (8.2 gigawatts) and Ireland (3.1 gigawatts) [1, 46]. Numerous strategies have recently been put up to cut down on the energy usage for blockchain technology and the resulting carbon footprint. One of these is switching from PoW validation to PoS (proof of stake). This more recent blockchain consensus algorithm has been suggested to solve PoW's scalability and cost issues. Secondly, creating blockchains that operate distinctly from those that use such large amounts of energy, lastly, concentrate on environmentally friendly methods to mine Bitcoin, such as solar or wind power. Assaults like 51% and denial-of-service (DoS) attacks impact blockchain technology. A 51% assault is one of the simplest methods for breaking the blockchain's security since it takes advantage of the consensus algorithm's legal purpose. In a PoW blockchain, the branch with the most work behind it wins in the event of a divergent blockchain. Hence the state of the blockchain is decided by a majority vote. A PoW blockchain is controlled by an attacker if they own 51% of its compute resources [1]. DoS attacks target the network's bottlenecks in conventional, centralized networks. Despite being decentralized and lacking bottlenecks by design, DoS attacks can be successful against blockchains. A DoS attack depends on blockchain and the locations of its operational bottlenecks [1].

4 Blockchain Technology in Healthcare

Blockchain is a novel technology that is employed to generate innovative solutions in a variety of industries including healthcare. Blockchain is a decentralized hyperledger that aids in the copying every activity or digital event preserves and distributes. The large number of persons involved confirms every transaction. Every transaction record is stored there [15]. A blockchain system is leveraged in the healthcare ecosystem to aid in keeping and sharing patient medical information among doctors, drug companies, diagnostic labs, and hospitals. Blockchain applications can precisely identify severe issues, including possibly deadly ones, in the medical sector. In the healthcare industry, it can enrich the privacy, translucence of shared medical records, and efficiency. Medical organizations can obtain perception and progress the investigation of patient information with the sustainable application of blockchain technology. Different blockchain systems can enhance the superiority of the services accessible to the medical industry. Blockchain and IoT technologies are coupled to allow healthcare institutions to maintain records effectively and accurately, which is necessary [1]. The procedure includes many parts, from the point at which IoT collects real-time patient data to the point at which a suitable medicine is given to ensure the patient's happiness. The blockchain is utilized in [1, 47] to maintain patients' medical records.

When blockchain technology's feasibility is considered, service-related industries such as healthcare are starting to transform and adapt to combine blockchain technology into their existing condition. By 2025, industry value is anticipated to grow by over $176 billion by 2025 and reach $3.1 trillion by 2030 [48]. According to reports from the World Economic Forum, blockchain technology will store 10% of the global GDP [49]. The worldwide blockchain industry's growth rate is expected to increase by 71.46% between 2017 and 2022, reaching $4.401 billion by 2022, up from $0.297 billion in 2017 [50]. These estimates do not surprise blockchain start-up investors, but the potential of blockchain remains to be discovered throughout society. Nonetheless, food procurement, mining, minerals, and banking are already using this technology [50, 51]. It is now being discussed as a potential application in healthcare to transform medical information systems [52] and payment systems [53, 54]. Researchers and professionals in the healthcare domain face scattered data, improper interactions, and clinical processes with incomplete components due to supplier-specific and unsuitable health systems. Furthermore, the confusion about transferring clinical and financial information impacts health IT systems. As a result, they have critical flaws in terms of privacy and security. Providing customized patient treatment is challenging due to all of these considerations [52]. The use of blockchain in the medical sector provides advantageous solutions for safeguarding stakeholder interactions, delivering clinical reports quickly, and combining various types of individual health records of people on a secure infrastructure.

4.1 A New Intelligent Healthcare System

Every organization or facility must have a system for the patient. This system must include a database to store the many distinct data sets needed by the institution and its stakeholders. These databases are mostly decentralized, containing several other databases working together. Decentralized databases help information management, notably in the healthcare sector [1, 55, 56]. For example, when several users submit scientific data, the database might grow perplexed and convoluted.

Blockchain technology may supply a remedy to certain knowledge areas, such as elderly treatment or chronic illnesses. Large healthcare databases, conflicting IT interfaces, changes in communication standards, and incompatibility of information processes among various therapists, medical specialists, practitioners, physicians, research labs, and healthcare centers, etc. are just a few of the significant issues that arise [1].

4.2 Improving the Privacy of Patients' Medical Records

Health information generated by patients and information stored are highly crucial [1, 57]. Wearable gadgets, such as trackers, built-in body chips, smartwatches, and fitness bands that help monitor patients' data, have been developed due to technological advancements in the medical sector. The flow of patient-generated information has grown due to these wearable technologies. The difficulties posed by the growth in healthcare data are addressed by blockchain technology [1, 56].

Health Bank's blockchain technologies offer Swiss start-ups in digital health. The following are the wide range of facilities to its customers [1]:

1. Management of patient transactions and the sharing of medical information
2. Retrieving patients' confidential information, such as lifestyle habits, health history, consumed medicines, eating habits, sleep patterns, heart rate, SpO2, and other vitals
3. Preservation and accessibility of information for medical research
4. Storage and management of information in a secure location

4.3 Healthcare Services Provided by Blockchain

The blockchain-based medical approach comprises the patient and clinical data generated and stored as digital resources and accessed over blockchain infrastructure. It is accessible under secure underlying architecture, seamless data usage, and exchange across health organizations and vendors. Blockchain applications in healthcare are health analytics, smart contracts, and a network of medical technologies that can use the complete history of timestamped records for users in terms of services. The following medical healthcare services are provided by the blockchain infrastructure framework [31]:

1. *Services provided by the doctors.* The suggested blockchain-based medical care approach improves clinical assistance provided by physicians and other healthcare providers. Computerized cost estimation via trade networks, such as online transportation services, contributes to translucent invoicing. Payments using coins based on blockchain offer an efficient payment method.

2. *Notary services.* A notary process for the digital transformation of vital records such as identity cards, passports, and insurance is one of the components employed for blockchain technology. In the proposed system, records such as examination marks, evidence of insurance, therapy, and medication are encrypted and validated in seconds rather than hours or days using traditional technology [32].

3. *Private blockchain technology.* Permission is required to view the data on a private blockchain. Therefore, patients grant access to their medical records to their doctors and other parties. Based on business agreements, different permit levels can be described to manage who can view documents, alter them, and have supreme control in the system. In healthcare, the mixture of safety and flexibility is preferable to the transparent public blockchain.

4. *Hospital asset supply chain management.* Blockchain technology can even be leveraged in supply chain control systems to handle hospital products and control purchase-selling mechanisms for all challenging clinical aids.

5. *Smart contract technology.* It could reduce insurance scams at all levels, from corporate fund management to specific claims, at a reduced cost because the system has less load to manage and process. It primarily performs as a legal mechanism, enabling real-time data exchange with the blockchain architecture to design, store, and implement agreements.

6. *Private genomic data.* Blockchain technology enables the collection of DNA records from various agencies and authorities without the requirement for a centralized database and the backup of a person's DNA to spread individual genetic material to multiple systems worldwide.

7. *Confidential medical record storage and access.* Clinical information enables seamless transmission and data exchange among healthcare communities and application merchandisers.

5 Quantum Computing in Healthcare

A promising computer method defined on quantum physics and associated exceptional circumstances is known as quantum computing (QC). It is a stunning combination of computational modeling, mathematics, physics, and computer science. Controlling the behavior of microscopic physical entities like atoms, photons, electrons, and other minute particles surpasses traditional computers in terms of low energy consumption, rapid speed, and processing capacity. Since the quantum approach is a more expansive physics model than classical physics, it adds to a universal computer system, quantum computing, which can solve issues that classical computing cannot [15]. In contrast to traditional systems, which store and process data using binary bits 0 and 1, quantum computing uses its quantum bits ("Qubits"). Although quantum computing can rapidly infiltrate current cryptography methods, the most extraordinary supercomputer currently known requires thousands of years to complete [15].

The future use of quantum computing in the medical sector may improve the data processing capacities of various stakeholders such as therapists, medical specialists, practitioners, physicians, research labs, and healthcare centers, which could be necessary for providing uninterrupted service in a changing environment. Executives should, however, account for (i) the current system's issues, that is, the matter of concern; (ii) the amount of information generated by the system; and (iii) the role in the medical sector (pharmaceutical or insurance company) [58]. Professionals in the medical industry, like clinicians, management staff, and insurance company support staff, must assess the risks of using quantum computing. In silico medical difficulties can be combined with quantum computing to create advanced virtual medical studies. The benefit of quantum computing is saving lives in danger in the conventional environment of clinical trials in the healthcare industry. Organizations and experts in healthcare insurance can benefit from precise risk modeling and provide a price when determining an individual's needs and health conditions [58]. Interindustry and interdepartmental coordination can also be presented with the integration of quantum computing to provide the most appropriate experience to a client in the medical industry. After meeting all requirements, healthcare organizations can also accept auto payments and provide funds as needed. For pharmaceutical industry experts in drug production, composition, and scalability, quantum computing can be highly profitable. The medical sector can concentrate on the patient while maintaining the highest level of security for medical information generated by electronic health records or clinical studies [58].

Although quantum computing will be capable of breaking some of the encryption methods used today, it is anticipated that they will develop tamperproof alternatives. Such tiny computers cannot use logic gates, semiconductors, and ICs. As a result, bits of atoms, protons, electrons, and ions are employed with metadata about their rotation and state. To create other combinations, they can be combined. As a result, they may operate concurrently and effectively use memory, enhancing their power. Since quantum computing is the only computer model that rejects the Church-Turing hypothesis, it can utilize the available systems far more effectively [15]. The essential building block of quantum theory is the qubit, which represents fundamental particles like atoms, protons, neutrons, electrons, and other minute particles as computer memory with their control schemes functioning as computer processors. It can be set to a weight of 0, 1, or both at once. Its processing power is a million times greater than the most sophisticated and powerful available today. Qubit production and management is a big task in engineering. The computing strength of quantum computers arrives from their digital and analog nature. Due to their analog character, there is no distortion limit for quantum gates; however, their digital characteristics provide a standard for overcoming this critical shortcoming [15]. Therefore, the representations and logic gates employed in traditional systems are meaningless in quantum computing. Mainly traditional computing rules can be employed in quantum computing. Nevertheless, this algorithm needs an extraordinary approach to eliminate processing variations and noise. It also needs its method for fixing design flaws and troubleshooting problems.

The three fundamental characteristics of quantum computing are entanglement, interference, and superposition [15, 59]. Entanglement is one of the critical features of quantum computing that guides the intimate connection between two particles or quantum bits. Qubits are connected in an exact instantaneous relationship even if they are split by excessive length, such as at opposite ends of the universe. They are coupled with one another or defined by one another. The quantum computing system simultaneously saves data in two states. In quantum computing, interference is equivalent to wave interference in classic physics. Wave interference results from the collision of two waves in the same space. However, imagine that all the waves are pointing in the same path. At that point, constructive or destructive interference happens as the generation of standing waves with individual amplitudes counted or a resultant wave with their amplitudes swabbed out, respectively. The resultant wave may be more or lesser than the initial wave, depending on the type of interference. A quantum technique's ability to live simultaneously in two separate locations or structures is known as superposition in quantum computing. It is distinct from its traditional companions, with binary limitations, and enables remarkable parallel processing at a fast speed [15].

5.1 Uses of Quantum Computing

There are many uses for quantum computing. Cybersecurity [60, 61], healthcare [62], artificial intelligence [63], weather forecasting, logistics optimization, and financial modeling are some of the critical uses of quantum computing. The Internet security environment has grown highly vulnerable due to the surge in cyberattacks that occur daily around the globe [64–66]. Although organizations are putting in place what is necessary for security standards, the process for traditional digital devices has grown complicated and impossible to use [15].

1. *Cybersecurity.* Large-scale quantum systems will enhance processing capacity, forming novel prospects for enhancing cybersecurity. Cybersecurity from the quantum era will be able to recognize and stop assaults from that era before they cause harm. However, it is a double-edged blade because quantum computing also creates new security flaws, such as the ability to quickly solve the complex mathematical equations that underlie some forms of encryption.
2. *Healthcare.* Integration of quantum computing with classical computing is expected to have considerable advantages over conventional computing alone in the healthcare sector. An extremely desirable set of skills, outstanding IT designs, a particular form of understanding, and inventive business plans are all demanded of quantum computing. In addition, security is impacted by technology, which is a subject of significant importance for the medical industry due to the medical industry's responsibilities and challenges concerning data security and privacy.
3. *Artificial intelligence.* AI and quantum computing are both game changers. Quantum computing is a necessary aspect for making significant advances in AI

factor. While generating practical applications on traditional computers, artificial intelligence is limited by its processing capability. Quantum computing may give artificial intelligence a boost in processing power that will let it tackle more complex problems in various fields.

4. *Financial modeling.* Companies using quantum computing benefit significantly from it. In particular, financial companies will be well prepared to evaluate large or unstructured information volumes. For example, banks might improve their judgments and customer service by making more timely or relevant offers. Quantum computers are showing promise in applications where programs are determined by actual information streams, such as livestock amounts, that include unexpectedly high noise.

5. *Logistics optimization.* The logistics sector could benefit significantly from quantum computing. Quantum computers would speed up devices using AI and machine learning, enhancing current CPUs. Quantum computers are showing promise in applications where programs are driven by actual information, such as livestock amounts, including unexpectedly high noise.

6. *Weather prediction.* On a native and worldwide level, quantum computing might aid to predict complex or accurate alerts of disastrous weather circumstances, thereby reducing yearly property damage and significantly saving lives. Quantum computing offers the ability to proceed with conventional mathematical approaches to improve tracking or forecasting weather circumstances by handling vast volumes of data with various variables fast or efficiently using the computational capability of qubits [15].

Pharmaceutical development and formulation are the most challenging jobs in quantum computing. Medicines are typically developed via costly, risky, or time-consuming trial and error. Quantum computing could be a helpful technique for examining the effects of medications on individuals, potentially saving a significant amount of time and money for pharmaceutical organizations. Since new emergent technologies have touched on almost every aspect of modern life, deep learning and AI are two of the most challenging cases [15]. Many problematic computational problems that take years to solve on traditional devices can be solved more quickly with quantum computing. Accounting professionals oversee enormous amounts of money, so even a slight modification in the projected return can have a significant impact. Another application is algorithmic investing, which is advantageous, primarily in high-volume trades, and involves a computer carrying out detailed algorithms to automatically launch stock trading based on market conditions. A cutting-edge, effective algorithm called quantum annealing may one day outperform conventional computers. On the other side, universal quantum computing is ready to handle any computing challenge but has yet to be commercially available [15].

5.2   Pros of Quantum Computing Based Approaches for Healthcare

When used in healthcare, quantum computing can [15]

1. Support quantum blockchain-based scenarios.
2. Enable innovative healthcare scenarios for better patient-centric system design and patient handling.
3. Improve security and enhance healthcare system for all stakeholders such as therapists, doctors, etc.
4. Speed up data availability and processing for authenticated clients.
5. Strengthen the healthcare environment security against a variety of real-time attacks.

5.3   Cons of Quantum Computing Based Approaches for Healthcare

Using quantum computing for healthcare systems faces several significant difficulties [15]. Some of them are as follows:

1. Quantum computing may not be ecologically friendly when extensively leveraged.
2. Deploying quantum computing possessions for the healthcare ecosystem in a real-world scenario is difficult particularly in underdeveloped countries.
3. Building the necessary infrastructure for real-time quantum computing applications is challenging, especially for the healthcare industry. Additionally, it is a pricey solution even though there are tools (such as Qiskit, Silq, etc.) for small-scale computing solutions. More Qibit support machines are needed in order to accomplish large-scale integration.

5.4   Quantum Cloud as a Service

Major cloud service providers are using serverless computing as a standard to efficiently deliver cloud services to end users [67, 68]. As edge computing becomes more prevalent, serverless computing can process work opportunities more quickly at runtime [69, 70]. It also provides services to end users established on the aids used by various Internet of Things (IoT) applications [71], like healthcare, smart cities, farming, and weather forecasting [72–74]. The computation speed and security issue for Serverless edge computing are sometimes faced due to big data processing [75–77]. Somewhat of data processing at cloud data centers, edge computing is a technique for handling data of IoT applications that are close to edge devices

[69, 74]. Computing close to the network's logical edge reduces latency and reaction time [41], but more processing power is needed as data creation grows daily [71, 78]. The cloud computing execution model-based service delivery to the end users on resource usage provided by IoT apps rather than pay-per-use [70, 78]. To manage resources and scale automatically, FaaS accomplishes the separation of the server into independent functions [38, 68].

6 Quantum and Blockchain in Healthcare

Blockchain technology uses conventional cryptographic operations to achieve security. Most of these functions are computationally secure, which implies that to breach them requires significant processing power that is not always available. These technologies will be impacted by the development of the quantum computer, which will make it possible to decrypt data encrypted with conventional encryption techniques. A quantum system can compromise the computational security of these operations. In contrast to traditional computers, quantum computers may effectively process information by taking advantage of peculiar quantum features, including superposition and quantum entanglement [1, 79–81]. As a result, it is expected that implementing quantum technologies in an intelligent environment would lead to innovations and accomplishments that are still unmatched by their classical counterparts. These enhancements include more affirmed security, quick computation, and little storage usage [1, 82–84].

Current cryptography foundations are seriously threatened by quantum computing [1, 85]. Within a few decades, according to current predictions, quantum computation will be potent sufficient to circumvent widely utilized privacy and security measures [1, 86, 87]. As a result, new technologies and gadgets must design for the environment of quantum computation and the ensuing cyberattacks. Future blockchain implementations must also be ready for quantum computing, as any flaw might allow for altering the database and threaten the whole organization's integrity. At last, legacy infrastructure will be exposed unless existing public fundamental cryptography architecture is updated to be quantum-resistant. With the key obtained via quantum key distribution systems, recently proven quantum blockchain infrastructures are based on information-theoretic safe authentication. However, a pairwise connection between users is necessary for such a configuration [1, 88]. Using quantum-secured direct contact for N clients with authentication or quantum digital signatures is another method for assuring quantum security. It is crucial to comprehend the quantum network structure restrictions and the number of false (unreliable) systems in these designs. The standard family of broadcast protocols can create quantum-secured distributed information methods once the authentication/signature operations are finished [1].

Aspects of healthcare, from detection and therapy to data storage and communication, could be impacted by innovations established on the rules of quantum mechanics. The fundamentals of quantum blockchain technology will increase medical data security and stop data leaks. Furthermore, by adopting methods like quantum computers and laser microscopy, which depend on the rules of quantum mechanics, we can sequence DNA more quickly and address other significant data issues in the healthcare field. This opens the door to individualized therapy that considers each person's genetic composition [1]. Edge serverless services that are secure and dependable can be offered while simultaneously enhancing security and computation speed using quantum computing and blockchain. Blockchain [39, 40] and quantum computing [89] are both the latest technology that can be leveraged to deliver high-speed computation and security [41]. Additionally, a safe and trustworthy theoretical ideal for serverless edge computation is required so that artificial intelligence (AI) can be used to deliver effective service [78]. Two key ideas from quantum physics, superposition, and entanglement are used to execute computations in quantum computing [89]. The serverless computing paradigm, now used by the edge computational paradigm to deliver operations as a service, needs quantum computing to handle extensive computation for load balancing and dynamic provisioning [38, 41].

6.1 *Importance of Quantum Blockchain*

Therefore, preserving participant security and anonymity is challenging for conventional healthcare infrastructure [90]. Blockchain has developed into a medium that increases the healthcare scheme's effectiveness while protecting all stakeholders' privacy. In this research, motivated by their findings, the authors [91] integrate quantum computing into the standard encryption system after looking at various security techniques used to protect medical records. As medical knowledge develops, using EMR systems to enhance the effectiveness and dependability of healthcare has become commonplace. Sharing problems arise when electronic medical record systems are housed independently in medical and healthcare institutions [92]. Additionally, highly delicate EMRs are open to manipulation and exploitation, posing privacy and security threats. The most recent developments of healthcare 4.0 integrating the IoT components [93] and cloud services to monitor clinical operations virtually have caught intellectuals' attention from a philosophical perspective.

7 Limitations and Scope for Future Research

When appropriately used, blockchain offers a reliable solution to problems with specific healthcare applications, such as real-time updates, security, accessibility, interoperability, integrity, privacy, sharing, and medical data. Blockchain, however, has constraints. Despite the benefits of using blockchain, substantial research obstacles existed before its development and deployment in the healthcare sector, necessitating its further study.

7.1 Challenges Related to Security

Blockchain technology has several distinct security faults in its application. Problems with the conventional agreement process utilized to validate the transactions are frequently connected to blockchain security concerns. Some security flaws are the 51% attack, block withholding, eclipse, selfish mining, block discarding, difficulty in rising, transaction malleability, and DDoS attacks. The distributed blockchain technique's agreement method cannot preclude these security risks. Because of the high resources needed, theoretical analysis cannot resolve the threats. The architecture of consensus methods is not particularly important for addressing these security issues. Due to the blockchain network's public nature, blockchain application introduces another possible vulnerability, pseudo-anonymity, where transactions can be tracked to discover physical identities or other data.

7.2 Challenges Related to Privacy

Users' or patients' privacy is not respected by the present secure communication architectures of EHR, as evidenced by noise in the data requester summary or the transfer system releasing all the information without the client's consent. To provide individualized services, the requester needs exact patient data if the current EHR systems are blockchain-based. Proposing a system that leverages cryptography algorithms for the privacy of information on blockchain-based EHRs is the main problem in ensuring the confidentiality of patient data. Due to this characteristic, it is challenging to determine any specific user using his current account number. Defects in protecting patient privacy data should be addressed in any framework of a similar nature. Integrating blockchain-based systems within EHRs demands increased processing capability and takes time to accomplish every operation, so patients should submit their data efficiently. Second, multiple processes are necessary to authenticate the sincere patient before joining a new system to the network, which is what new clients require.

7.3 Challenges Related to Restrictions and Latency

Integrating blockchain with applications in the medical field that respond to data and events in real time may be difficult because most blockchain technologies require much time to reach desired consensus and complete transactions that need to be completed. A blockchain needs time to process transactions when there is transaction latency. Contrarily, most conventional database infrastructures only need a few beats to approve an entry. Due to throughput limitations, EHRs and RPM in the IoT are built on a blockchain. Procedures often require enormous processing amounts of transactions per second, posing a potential difficulty for blockchains.

7.4   Challenges Related to Blockchain Size

Blockchains grow more complex as more devices execute transactions, such as EHRs and IoT-RPM, necessitating the deployment of more powerful miners. The old IoMT devices cannot handle even the smallest blockchains because of resource limitations. Therefore, it is important to research other compression techniques in the blockchain, such as mini-blockchains.

7.5   Challenges Related to Computing Power Limitations

Blockchain data from IoMT devices are frequently computationally constrained, making it possible that no encryption techniques will be required. Cryptosystems in devices with limited computational resources, such as memory and processing power, handle sensor and actuator protection in many health-related applications. In other words, they are met with current, safe public-key cryptography techniques. Most blockchains use public-key cryptosystems, which have efficiency and security difficulties, making it challenging to choose practical cryptography. Blockchain cryptosystems must be familiar with the threat posed by post-quantum computing and seek energy-efficient quantum-safe methods to maintain data security for prolonged periods.

7.6   Challenges Related to Storage

Restrictive systems that transmit information to the network may run into issues because a blockchain needs much storage to record complete network transactions. Blockchain can guarantee that the distributed, large-scale EHR data is neither altered, unforgeable, nor falsifiable, but it may suffer from the storage requirements of such data.

7.7   Challenges Related to Scalability

Blockchain's infrastructure could need to handle computing requirements due to the issue of increased number of system users. Because the computing power of many smart devices or sensors is less than that of a typical computer, the problem becomes more challenging.

7.8   Challenges Related to Interoperability and Standardization

Healthcare application interoperability needs to be improved by a lack of information collection, exchange, and analysis frameworks. The management of the current EHR systems relies on offline architecture and centralized local databases, whereas cloud-based blockchain technology is decentralized. As a result, if healthcare organizations use blockchain technology, an effective electronic health record system that can promote communication and cooperation between the medical and scientific communities is a prerequisite. Many technical issues need to be resolved regarding the moved EHR information.

Data processing in the healthcare ecosystem is primarily manual in large hospitals, huge drug industries, and pharmacy shops. The external parties that affect an organization's operations are excluded from organizational information processing theory. For instance, the functions of a hospital are directly impacted by the health insurance company, and these activities also affect the pharmaceutical industry, which impacts the patient. Building a quantum computing ecosystem requires a fundamental computing infrastructure and operations volume. The future study may consider a sizable sample of well-known actors from the hospital, pharmaceutical, and health insurance sectors to grasp the geographic advancements. Although the findings do not specifically mention the importance of machine learning and artificial intelligence, it is clear how beneficial quantum computing might be for the medical sector. Therefore, future research can examine how artificial intelligence and machine learning enable quantum computing in the medical industry.

Future research should also account for how quantum computing affects the healthcare sector in a post-COVID era and how it is essential in bringing together the various players, including insurance agents, hospitals, pharmaceutical firms, payers, and patients. For a particular group or kind of subindustry, it is possible to identify standard quantum computing techniques, for instance, the size, industry, and type of data that a hospital, pharmaceutical, or health insurance company produces daily.

8 Conclusion

Blockchain technology in healthcare has been introduced to enhance extensive data analysis, administration, and security. Considering the sensitivity and real-time processing requirement of patients' clinical data, the agreement algorithm, working medium, and blockchain type need to be focused on ensuring proper blockchain access to safeguard the patient's personal information. It has been observed that while implementing blockchain and IoHT, healthcare executives are accelerating the adoption process of the leading technologies, and now it is the time to value artificial intelligence, which collects and extracts insights from massive amounts of data by identifying patterns and correlations. In addition to artificial intelligence, hybrid clouds are suggested as a solid blockchain and IoHT foundation. These new AI-based hybrid clouds at client sites deliver highly scalable cloud services and applications while maintaining medical information behind the firewalls to meet organizational and regulatory requirements. This technology enables users to access large amounts of data at any time by storing it orderly and securely. It is possible to hide data in a quantum blockchain while ensuring its security and accessibility. Quantum blockchain technology made it feasible to process user data faster while keeping its virtue by utilizing both quantum computing and blockchain technologies.

References

1. Farouk, A., Alahmadi, A., Ghose, S., & Mashatan, A. (2020). Blockchain platform for industrial healthcare: Vision and future opportunities. In *Computer communications* (Vol. 154, pp. 223–235). Elsevier B.V. https://doi.org/10.1016/j.comcom.2020.02.058
2. Ivanteev, A., Ilin, I., & Iliashenko, V. (2020). Possibilities of blockchain technology application for the health care system. In *IOP conference series: Materials science and engineering* (Vol. 940(1)). IOP Publishing. https://doi.org/10.1088/1757-899X/940/1/012008
3. Liang, X., Zhao, J., Shetty, S., Liu, J., & Li, D. (2017). Integrating blockchain for data sharing and collaboration in mobile healthcare applications. In *2017 Presented at: Personal, indoor, Mobile radio communications (PIMRC), IEEE 28th annual international symposium; Montreal* (pp. 1–5).
4. Arora, S., Lamba, D., & V. (2020). A study of technologies to further research in health care data security in medical report using block chain. *International Journal of Advanced Engineering Research and Science, 7*(6), 248–252. https://doi.org/10.22161/ijaers.76.31
5. Christidis, K., & Devetsikiotis, M. (2016). Blockchains and smart contracts for the Internet of Things. *IEEE Access, 4*, 2292–2303.
6. Liu, D., Alahmadi, A., Ni, J., Lin, X., & Shen, X. (2019). Anonymous reputation system for IIoT-enabled retail marketing atop PoS blockchain. *IEEE Transactions on Industrial Informatics, 15*(6), 3527–3537.
7. Ali, A., Rahouti, M., Latif, S., Kanhere, S., Singh, J., Janjua, U., & Crowcroft, J. (2019). Blockchain and the future of the internet: A comprehensive review. *arXiv preprint arXiv: 1904.00733*.
8. Islam, S. R., Kwak, D., Kabir, M. H., Hossain, M., & Kwak, K. S. (2015). The Internet of Things for health care: A comprehensive survey. *IEEE Access, 3*, 678–708.

9. Kranz, M. (2016). *Building the Internet of Things: Implement new business models, disrupt competitors, transform your industry*. John Wiley & Sons.
10. Atzori, L., Iera, A., & Morabito, G. (2010). The Internet of Things: A survey. *Computer Networks, 54*(15), 2787–2805.
11. Zyskind, G., & Nathan, O. (2015). Decentralizing privacy: Using blockchain to protect personal data. In *2015 IEEE security and privacy workshops* (pp. 180–184). IEEE.
12. Mougayar, W. (2016). *The business blockchain: Promise, practice, and application of the next internet technology*. John Wiley & Sons.
13. McGhin, T., Choo, K. K. R., Liu, C. Z., & He, D. (2019). Blockchain in healthcare applications: Research challenges and opportunities. *Journal of Network and Computer Applications, 135*, 62–75.
14. Mettler, M. (2016). Blockchain technology in healthcare: The revolution starts here. In *2016 IEEE 18th international conference on e-health networking, applications and services, Healthcom* (pp. 1–3).
15. Kaushik, K., & Kumar, A. (2023). Demystifying quantum blockchain for healthcare. *Security and Privacy, 6*(3), e284.
16. Hussien, H. M., Yasin, S. M., Udzir, S. N. I., Zaidan, A. A., & Zaidan, B. B. (2019). A systematic review for enabling of develop a blockchain technology in healthcare application: Taxonomy, substantially analysis, motivations, challenges, recommendations and future direction. *Journal of Medical Systems, 43*(10), 1–35. https://doi.org/10.1007/s10916-019-1445-8
17. Casino, F., Dasaklis, T. K., & Patsakis, C. (2019). A systematic literature review of blockchain-based applications: Current status, classification and open issues. In *Telematics and informatics* (Vol. 36, pp. 55–81). Elsevier Ltd. https://doi.org/10.1016/j.tele.2018.11.006
18. Bhattacharya, P., Tanwar, S., Bodkhe, U., Tyagi, S., & Kumar, N. (2021). BinDaaS: Blockchain-based deep-learning as-a-service in healthcare 4.0 applications. *IEEE Transactions on Network Science and Engineering, 8*(2), 1242–1255. https://doi.org/10.1109/TNSE.2019.2961932
19. Mistry, I., Tanwar, S., Tyagi, S., & Kumar, N. (2020). Blockchain for 5Genabled IoT for industrial automation: A systematic review, solutions, and challenges. *Mechanical Systems and Signal Processing, 135*, 1–21.
20. Kabra, N., Bhattacharya, P., Tanwar, S., & Tyagi, S. (2020). Mudrachain: Blockchain-based framework for automated cheque clearance in financial institutions. *Future Generation Computer Systems, 102*, 574–587.
21. Srivastava, A., Bhattacharya, P., Singh, A., Mathur, A., Prakash, O., & Pradhan, R. (2018). A distributed credit transfer educational framework based on blockchain. In *Proc. 2nd int. conf. Advances Comput., control Commun. Technol., Allahabad, India* (pp. 54–59).
22. Bhattacharya, P., Tanwar, S., Shah, R., & Ladha, A. (2020). Mobile edge computing- enabled blockchain framework—A survey. In P. K. Singh, A. K. Kar, Y. Singh, M. H. Kolekar, & S. Tanwar (Eds.), *Proceedings of International Conference on Recent Innovations in Computing* (pp. 797–809). Springer.
23. Bodkhe, U., Bhattacharya, P., Tanwar, S., Tyagi, S., Kumar, N., & Obaidat, M. (2019). Blohost: Blockchain enabled smart tourism and hospitality management. In *Proceedings of international conference on computer, information and telecommunication systems, Beijing, China* (pp. 1–5).
24. Healthcare data management meet blockchain. Accessed 28 Mar 2019. [Online]. Available: https://steemit.com/healthcare/@robmenzies/healthcare-datamanagement-meet-blockchain
25. Vora, J., et al. (2018). Bheem: A blockchain-based framework for securing electronic health records. In *Proceedings of IEEE Globecom Workshops, Abu-Dhabi, UAE* (pp. 1–6).
26. Li, C.-Y., Chen, X.-B., Chen, Y.-L., Hou, Y.-Y., & Li, J. (2019). A new lattice based signature scheme in post-quantum blockchain network. *IEEE Access, 7*, 2026–2033.
27. Salah, K., Rehman, M. H. U., Nizamuddin, N., & Al-Fuqaha, A. (2019). Blockchain for AI: Review and open research challenges. *IEEE Access, 7*, 10127–10149.
28. Murphy, K. P. (2012). *Machine learning: A probabilistic perspective*. MIT Press.

29. Domingos, P. M. (2012). A few useful things to know about machine learning. *Communications of the ACM, 55*(10), 78–87.
30. Yaqoob, I., Salah, K., Jayaraman, R., & Al-Hammadi, Y. (2022). Blockchain for healthcare data management: Opportunities, challenges, and future recommendations. *Neural Computing and Applications, 34*(14), 11475–11490. https://doi.org/10.1007/s00521-020-05519-w
31. Gökalp, E., Gökalp, M. O., Çoban, S., & Eren, P. E. (2018). Analysing opportunities and challenges of integrated blockchain technologies in healthcare. In S. Wrycza & J. Maślankowski (Eds.), *Information systems: Research, development, applications, education. SIGSAND/PLAIS 2018* (Lecture notes in business information processing) (Vol. 333). Springer. https://doi.org/10.1007/978-3-030-00060-8_13
32. Swan, M. (2015). *Blockchain: Blueprint for a new economy*. O'Reilly Media Inc..
33. Hasselgren, A., Kralevska, K., Gligoroski, D., Pedersen, S. A., & Faxvaag, A. (2020). Blockchain in healthcare and health sciences—A scoping review. *International Journal of Medical Informatics, 134*, 104040.
34. Xie, J., Tang, H., Huang, T., Yu, F. R., Xie, R., Liu, J., & Liu, Y. (2019). A survey of blockchain technology applied to smart cities: Research issues and challenges. *IEEE Communications Surveys & Tutorials, 21*(3), 2794–2830.
35. Zheng, Z., Xie, S., Dai, H.-N., Chen, W., Chen, X., Weng, J., & Imran, M. (2020). An overview on smart contracts: Challenges, advances and platforms. *Future Generation Computer Systems, 105*, 475–491.
36. Gupta, S., Malhotra, V., & Singh, S. N. (2020). Securing IOT-driven remote healthcare data through blockchain. In *Advances in data and information sciences* (Lecture notes in networks and systems) (Vol. 94, pp. 47–56). Springer.
37. Ali, M. S., Vecchio, M., Pincheira, M., Dolui, K., Antonelli, F., & Rehmani, M. H. (2018). Applications of blockchains in the internet of things: A comprehensive survey. *IEEE Communications Surveys & Tutorials, 21*(2), 1676–1717.
38. Gill, S. S. (2021). Quantum and blockchain based serverless edge computing: A vision, model, new trends and future directions. *Internet Technology Letters, 7*, e265. https://doi.org/10.1002/itl2.275
39. Cui, L., Chen, Z., Yang, S., et al. (2020). A blockchain-based containerized edge computing platform for the internet of vehicles. *IEEE Internet of Things Journal, 8*(4), 2395–2408.
40. Zheng, Z., Xie, S., Dai, H. N., Chen, X., & Wang, H. (2018). Blockchain challenges and opportunities: A survey. *International Journal of Web and Grid Services, 14*(4), 352–375.
41. Tuli, S., Basumatary, N., Gill, S. S., et al. (2020). Healthfog: An ensemble deep learning based smart healthcare system for automatic diagnosis of heart diseases in integrated IoT and fog computing environments. *Future Generation Computer Systems, 104*, 187–200.
42. Sankar, L. S., Sindhu, M., & Sethumadhavan, M. (2017). Survey of consensus protocols on blockchain applications. In *2017 4th international conference on advanced computing and communication systems* (pp. 1–5). IEEE.
43. Xu, X., Weber, I., Staples, M., Zhu, L., Bosch, J., Bass, L., Rimba, P., et al. (2017). A taxonomy of blockchain-based systems for architecture design. In *2017 IEEE international conference on software architecture, ICSA* (pp. 243–252). IEEE.
44. Gatteschi, V., Lamberti, F., Demartini, C., Pranteda, C., & Santamaria, V. (2018). To blockchain or not to blockchain: That is the question. *IT Professional, 20*(2), 62–74.
45. Niranjanamurthy, M., Nithya, B. N., & Jagannatha, S. (2019). Analysis of blockchain technology: Pros, cons and SWOT. *Cluster Computing, 22*(6), 14743–14757.
46. De Vries, A. (2018). Bitcoin's growing energy problem. *Joule, 2*(5), 801–805.
47. Ekblaw, A., Azaria, A., Halamka, J. D., & Lippman, A. (2016). A case study for blockchain in healthcare: MedRec prototype for electronic health records and medical research data. In *Proceedings of IEEE open & big data conference* (Vol. 13, p. 13).
48. Gartner: Forecast: Blockchain Business Value, Worldwide, 2017–2030 (2017).
49. World Economic Forum: Deep Shift. Technology Tipping Points and Societal Impact (2015).
50. Research and Markets: Blockchain Market - Forecasts from 2017 to 2022 (2017).

51. ONC: Connecting health and care for the nation: a 10-year vision to achieve an interoperable health it infrastructure (2014).
52. Middleton, B., et al. (2013). Enhancing patient safety and quality of care by improving the usability of electronic health record systems: Recommendations from AMIA. *Journal of the American Medical Informatics Association, 20*(e1), e2–e8.
53. Mettler, M. (2016). Blockchain technology in healthcare: The revolution starts here. In *2016 IEEE 18th international conference on e-health networking, applications and services, Healthcom 2016* (pp. 1–3).
54. Basu, A., Subedi, P., & Kamal-Bahl, S. (2016). Financing a cure for diabetes in a multi payer environment. *Value in Health, 19*(6), 861–868.
55. Lu, Y. (2018). Blockchain and the related issues: A review of current research topics. *Journal of Management Analytics, 5*(4), 231–255.
56. Yue, X., Wang, H., Jin, D., Li, M., & Jiang, W. (2016). Healthcare data gateways: Found healthcare intelligence on blockchain with novel privacy risk control. *Journal of Medical Systems, 40*(10), 218.
57. Hölbl, M., Kompara, M., Kamišalić, A., & Nemec Zlatolas, L. (2018). A systematic review of the use of blockchain in healthcare. *Symmetry, 10*(10), 470.
58. Gupta, S., Modgil, S., Bhatt, P. C., Chiappetta Jabbour, C. J., & Kamble, S. (2022). Quantum computing led innovation for achieving a more sustainable Covid-19 healthcare industry. *Technovation, 120*, 102544. https://doi.org/10.1016/j.technovation.2022.102544
59. Marella, S. T., Parisa, H. S. K., & Parisa, K. (2020, October). Introduction to quantum computing. *Quantum Computing and Communications*. https://doi.org/10.5772/INTECHOPEN.94103
60. 8 Quantum computing applications you should know | built in. https://builtin.com/hardware/quantum-computing-applications. Accessed 18 May 2022.
61. Kaushik, K., & Dahiya, S. (2022). Scope and challenges of blockchain technology. *Lecture Notes in Electrical Engineering, 832*, 461–473. https://doi.org/10.1007/978-981-16-8248-3_38/COVER/
62. Kaushik, K., Dahiya, S., & Sharma, R. (2021). Internet of Things advancements in healthcare. *Internet of Things*, 19–32. https://doi.org/10.1201/9781003140443-2
63. Singh, M., Dhara, C., Kumar, A., Gill, S. S., & Uhlig, S. (2021). *Quantum artificial intelligence for the science of climate change*. https://doi.org/10.48550/arxiv.2108.10855
64. Top applications of quantum computing everyone should know about. https://analyticsindiamag.com/top-applications-of-quantum-computing-everyone-should-know-about/. Accessed 18 May 2022.
65. Chugh, N., Kumar, A., & Aggarwal, A. (2016). Security aspects of a RFID-sensor integrated low-powered devices for Internet-of-Things. In *2016 4th international conference on parallel, distributed and grid computing, PDGC 2016* (pp. 759–763). https://doi.org/10.1109/PDGC.2016.7913223
66. Vashisht, S., Gaba, S., Dahiya, S., & Kaushik, K. (2022). Security and privacy issues in IoT systems using blockchain. *Sustainable and Advanced Applications of Blockchain in Smart Computational Technologies*, 113–127. https://doi.org/10.1201/9781003193425-8
67. Gill, S. S., Tuli, S., Xu, M., et al. (2019). Transformative effects of IoT, blockchain and artificial intelligence on cloud computing: Evolution, vision, trends and open challenges. *Internet of Things, 8*, 100118.
68. Malla, S., & Christensen, K. (2020). HPC in the cloud: Performance comparison of function as a service (FaaS) vs infrastructure as a service (IaaS). *Internet Technology Letters, 3*(1), e137.
69. Tuli, S., Gill, S. S., Casale, G., & Jennings, N. R. (2020). iThermoFog: IoT-fog based automatic thermal profile creation for cloud data centers using artificial intelligence techniques. *Internet Technology Letters, 3*(5), e198.

70. Bansal, K., Mittal, K., Ahuja, G., Singh, A., & Gill, S. S. (2020). DeepBus: Machine learning based real time pothole detection system for smart transportation using IoT. *Internet Technology Letters, 3*(3), e156.

71. Cicconetti, C., Conti, M., & Passarella, A. (2020). A decentralized framework for serverless edge computing in the Internet of Things. *IEEE Transactions on Network and Service Management, 18*, 1–14. https://doi.org/10.1109/TNSM.2020.3023305

72. ShiW, C. J., Zhang, Q., Li, Y., & Xu, L. (2016). Edge computing: Vision and challenges. *IEEE Internet of Things Journal, 3*(5), 637–646.

73. Baldini, I., Castro, P., Chang, K., et al. (2017). Serverless computing: Current trends and open problems. In C. Sanjay, S. Gaurav, & B. Rajkumar (Eds.), *Research advances in cloud computing* (pp. 1–20). Springer.

74. Tuli, S., Tuli, S., Wander, G., et al. (2020). Next generation technologies for smart healthcare: Challenges, vision, model, trends and future directions. *Internet Technology Letters, 3*(2), e145.

75. Moret-Bonillo, V. (2015). Can artificial intelligence benefit from quantum computing? *Progress in Artificial Intelligence, 3*(2), 89–105.

76. Benedict, S. (2020). Serverless blockchain-enabled architecture for IoT societal applications. *IEEE Transactions on Computational Social Systems, 7*, 1146–1158.

77. Dunjko, V., & Briegel, H. J. (2018). Machine learning & artificial intelligence in the quantum domain: A review of recent progress. *Reports on Progress in Physics, 81*(7), 074001.

78. Aslanpour, M. S., Toosi, A., Cicconetti, C., et al. (2021). Serverless edge computing: Vision and challenges. In *Proceedings of the 19th Australasian symposium on parallel and distributed computing (AusPDC 2021)* (pp. 1–10).

79. Metwaly, A. F., Rashad, M. Z., Omara, F. A., & Megahed, A. A. (2014). Architecture of multicast centralized key management scheme using quantum key distribution and classical symmetric encryption. *European Physical Journal Special Topics, 223*(8), 1711–1728.

80. Farouk, A., Zakaria, M., Megahed, A., & Omara, F. A. (2015). A generalized architecture of quantum secure direct communication for N disjointed users with authentication. *Scientific Reports, 5*(1), 1–17.

81. Naseri, M., Raji, M. A., Hantehzadeh, M. R., Farouk, A., Boochani, A., & Solaymani, S. (2015). A scheme for secure quantum communication network with authentication using GHZ-like states and cluster states controlled teleportation. *Quantum Information Processing, 14*(11), 4279–4295.

82. Zhou, N. R., Li, J. F., Yu, Z. B., Gong, L. H., & Farouk, A. (2017). New quantum dialogue protocol based on continuous-variable two-mode squeezed vacuum states. *Quantum Information Processing, 16*(1), 4.

83. Farouk, A., Batle, J., Elhoseny, M., Naseri, M., Lone, M., Fedorov, A., Abdel-Aty, M., et al. (2018). Robust general N user authentication scheme in a centralized quantum communication network via generalized GHZ states. *Frontiers of Physics, 13*(2), 130306.

84. Abulkasim, H., Farouk, A., Alsuqaih, H., Hamdan, W., Hamad, S., & Ghose, S. (2018). Improving the security of quantum key agreement protocols with single photon in both polarization and spatial-mode degrees of freedom. *Quantum Information Processing, 17*(11), 316.

85. Shor, P. W. (1994). Algorithms for quantum computation: Discrete logarithms and factoring. In *Proceedings 35th annual symposium on foundations of computer science* (pp. 124–134).

86. Mosca, M. (2018). Cybersecurity in an era with quantum computers: Will we be ready? *IEEE Security and Privacy, 16*(5), 38–41.

87. Bauer, B., Wecker, D., Millis, A. J., Hastings, M. B., & Troyer, M. (2016). Hybrid quantum–classical approach to correlated materials. *Physical Review X, 6*(3), 031045.

88. Kiktenko, E. O., Pozhar, N. O., Anufriev, M. N., Trushechkin, A. S., Yunusov, R. R., Kurochkin, Y. V., Fedorov, A. K., et al. (2018). Quantum-secured blockchain. *Quantum Science and Technology, 3*(3), 035004.

89. Gill, S. S., Kumar, A., Singh, H., et al. (2020). Quantum computing: A taxonomy, systematic review and future directions. *arXiv preprint arXiv:2010.15559.*

90. Singh, K., Kaushik, K., Ahatsham, & Shahare, V. (2020). Role and impact of wearables in IoT healthcare. *Advances in Intelligent Systems and Computing, 1090*, 735–742. https://doi.org/10.1007/978-981-15-1480-7_67

91. Bhavin, M., Tanwar, S., Sharma, N., Tyagi, S., & Kumar, N. (2021). Blockchain and quantum blind signature-based hybrid scheme for healthcare 5.0 applications. *Journal of Information Security and Applications, 56*, 102673. https://doi.org/10.1016/J.JISA.2020.102673

92. Kumar, A., Krishnamurthi, R., Nayyar, A., Sharma, K., Grover, V., & Hossain, E. (2020). A novel smart healthcare design, simulation, and implementation using healthcare 4.0 processes. *IEEE Access, 8*, 118433–118471. https://doi.org/10.1109/ACCESS.2020.3004790

93. A. Kumar, C. Ottaviani, S. S. Gill, and R. Buyya, "Securing the future internet of things with post-quantum cryptography," Security and Privacy, vol. 5, no. 2, p. e200, Mar. 2022, doi: https://doi.org/10.1002/SPY2.200.

Innovative Solutions for Sustainability: Quantum and Blockchain Technologies

Ahmed Mateen Buttar, Nouman Arshad, and Muhammad Azeem Akbar

Abstract Quantum computing could revolutionize energy optimization. Quantum computers can improve energy prediction, waste reduction, and efficiency. Blockchain technology can make supply chains more transparent and sustainable. Tracking and validating the supply chain from raw materials to completed products ensures ethical and sustainable sourcing. Quantum cryptography secures data transmission using quantum mechanics. Its impenetrable security makes it excellent for data sharing in sustainability programs. Nonclassical computation is a component of quantum computing, and it has been demonstrated to be more effective than classical computation in solving some key issues like factoring big integers. This work frequently read stories asserting that blockchain will be threatened by quantum computing, which would alter its cryptographic foundation, and communication protocols, and impair the record's immutability. It is only a matter of time before threats to blockchain and cryptocurrency materialize because nation-state actors are purposefully building quantum computers that can be obtainable via the cloud and can break the present cryptography. In juxtaposition with today's conventional computers, quantum computers operate differently. Quantum computers conduct some types of computations that are more powerful than what our existing computers can do today by using subatomic phenomena like entanglement and superposition. Additionally, quantum simulation and quantum qubit algorithms are potential to enhance computations in drug discovery. In comparison with classical bits, qubits, the bits used in quantum computers, have two advantages: they can hold many states throughout computing, and two qubits can be entangled. Solutions that can safeguard data while producing an immutable record demonstrating data has not been altered are becoming more and more necessary. Due to its decentralized nature,

A. M. Buttar (✉) · N. Arshad
Department of Computer Science, University of Agriculture Faisalabad, Faisalabad, Pakistan

M. A. Akbar
Department of Software Engineering, Lappeenranta-Lahti University of Technology (LUT University), Lappeenranta, Finland
e-mail: azeem.akbar@lut.fi

© The Author(s), under exclusive license to Springer Nature
Switzerland AG 2024
S. Pulipeti et al. (eds.), *Quantum and Blockchain-based Next Generation
Sustainable Computing*, Contributions to Environmental Sciences & Innovative
Business Technology, https://doi.org/10.1007/978-3-031-58068-0_3

consensus system of checks and balances, and cryptographic foundation used to secure the data in a protected and validated state, blockchain is viewed as having a high level of security. Individuals and huge funds, it has developed into their financial asset class as a result of these factors. Using the most recent advances in cryogenic engineering, quantum science, industrial design, and systems engineering, we are now developing cutting-edge superconducting quantum processors integrated into classical computing operations. Blockchain technology, which underpins cryptocurrencies like Bitcoin, allows for the speedier clearing of funds as well as trading without the need for a traditional bank or credit institution. Quantum algorithm can improve data integrity, processing speed, and protection.

Keywords Hybrid encryption · Post-quantum signature · Quantum algorithm · Quantum cryptography · Quantum entanglement · Quantum gate · Quantum key · Quantum teleportation · Quantum state

1 Introduction

1.1 *Quantum Computation*

Quantum computing offers the potential for faster and more reliable computations compared to conventional computers. A QC originates from a quantum circuit with wires and basic quantum gates to transmit and operate quantum information, similar to how a conventional computer is constructed from an electrical circuit containing wires and logic gates [1]. By counting a circuit that teleports qubits, this work names some basic quantum gates in this section and shows example circuits demonstrating their use.

1.2 *Single-Qubit Gates*

Wires and logic gates make up typical computer circuits. While the logic gates digest data and convert it from one form to another, the wires are employed to transfer information across the circuit. Consider the standard single-bit logic gates, for example. The gate's mechanism is specified according to its truth table; the $0 \rightarrow 1$ states are interchanged and are the sole complex member of this class. The quantum gate behaves linearly, taking the following state as input

$$\alpha \, | \, 0 \rangle + \beta \, | \, 1 \rangle \tag{1}$$

to the corresponding condition where $|0|$ and $|1|$ have switched places:

$$\alpha \, | \, 1 \rangle + \beta \, | \, 0 \rangle \tag{2}$$

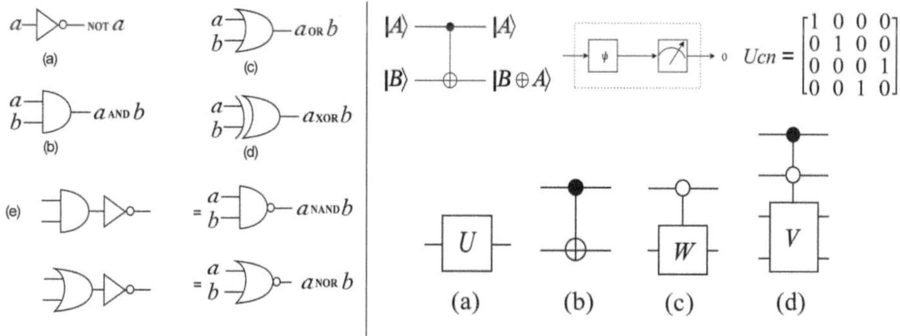

Fig. 1 Single and multiple-bit gates are shown on the left, and a common multi-qubit gate is shown on the right

1.3 Multi-Qubit Gates

Let's now reduce the complexity to several qubits. Figure 1 depicts the classical multiple-bit AND, OR, XOR (exclusive-OR), and gates for five personalities. The fact that slight meaning on bits may stand derived directly after the configuration of gates, which is referred to as a general gate, is a significant theoretical result [2]. The skillful gate is a multi-qubit quantum logic gate. The circuit is depicted in the higher correct turning of Fig. 1. The switch qubit is represented on the highest line, and the goal qubit is displayed on the bottom line. The gate's behavior can be classified as follows:

$$|00\rangle \rightarrow |00\rangle; |01\rangle \rightarrow |01\rangle; |10\rangle \rightarrow |11\rangle; |11\rangle \rightarrow |10\rangle \tag{3}$$

1.4 Measurement in Bases Other than the Computational Basis

The quantities of a single qubit in the state $\alpha|0\rangle + \beta|1\rangle$ are labeled as soft, representing the possible outcomes of zero or one. The qubit exists in a consistent state of $|0\rangle$ or $|1\rangle$, with possibilities $|\alpha|2$ and $|\beta|2$, respectively, as shown [3] in Fig. 1. Quantum mechanics allows for some degree of flexibility in the range of possible sizes, but it is not enough to accurately determine both the size and outcome of a single measurement.

Keep in mind that the states $|0\rangle$ and $|1\rangle$ represent just two of the numerous potential origin states for a qubit. The sets $|+\rangle \equiv (|0\rangle + |1\rangle)/\sqrt{2}$ and $|-\rangle \equiv (|0\rangle - |1\rangle)/\sqrt{2}$ are another potential option. The relationships between the states $|+\rangle$ and $|-\rangle$ can be used to repress the chance state $|\psi\rangle = \alpha |0\rangle + \beta |1\rangle$:

$$|\psi\rangle = \alpha\,|0\rangle + \beta\,|1\rangle = \alpha\frac{|+\rangle+|-\rangle}{\sqrt{2}} + \beta\frac{|+\rangle+|-\rangle}{\sqrt{2}} = \frac{\alpha+\beta}{\sqrt{2}}\,|+\rangle + \frac{\alpha-\beta}{\sqrt{2}}\,|-\rangle \qquad (4)$$

Additionally, if the states are orthonormal, it is possible to measure about the $|a\rangle$ $|b\rangle$ basis and obtain the results $|\alpha|^2$ and $|\beta|^2$ with the possibility (2). It is necessary to impose the orthonormality restriction so that in $|\alpha|^2 + |\beta|^2 = 1$ the possibilities are as expected.

1.5 Quantum Circuits

We've previously encountered a unique, modest quantum circuit. Agreements examine the principles of a quantum circuit in a little more detail. Figure 2 depicts a basic quantum circuit with three quantum gates and read the circuit from left to right; each circuit line represents a wire [4]. This wire may resemble an alternate to time or even a bodily atom like a photon, an atom of brightness traveling across space, rather than inexorably resembling a real wire. The two qubits' states are switched by the tour in Fig. 2, which achieves a straightforward yet valuable purpose [3]. The memo that the arrangement of gates has the succeeding arrangement foundation state a and state b demonstrates how this circuit achieves the swap process:

$$|a, b\rangle \rightarrow |a, a \oplus b\rangle \qquad (5)$$

$$\rightarrow |a \oplus (a \oplus b), a \oplus b\rangle = |b, a \oplus b\rangle \qquad (6)$$

$$\rightarrow |b, (a \oplus b) \oplus b\rangle = |b, a\rangle \qquad (7)$$

wherever Modulation 2 stays used for all accompaniments. Therefore, the circuit's conclusion involves trading the two-qubit states.

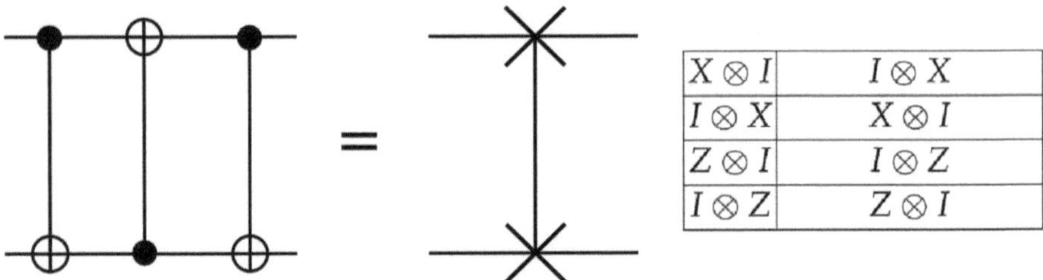

Fig. 2 Equivalent schematic symbol notation practical circuit that switches two qubits is provided

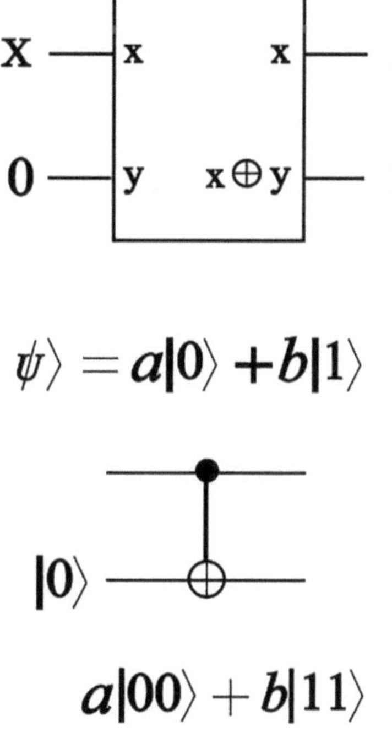

Fig. 3 Using both classical and quantum circuits

$$\psi\rangle = a|0\rangle + b|1\rangle$$

$$a|00\rangle + b|11\rangle$$

1.5.1 Qubit Copying Circuit

The gate is valuable as a depiction of one of the maximum significant characteristics of quantum info. Think of the replication task as a traditional bit. As shown in Fig. 3, this can be accomplished by utilizing a conventional gate that accepts and has been reset to zero [5]. Two bits are output, both of which are in the same state (x). Consider using the same procedure with a gate to replicate state $|\psi\rangle = a|0\rangle + b|1\rangle$. It is advisable to use the same procedure using a gate to replicate the state $|\psi\rangle = a|0\rangle + b|1\rangle$. The initial state of the two qubits can be expressed as follows:

$$\left[a|0\rangle + b|1\rangle\right]|0\rangle = a|00\rangle + b|10\rangle \tag{8}$$

That is what this circuit accomplishes also explain [2] in Fig. 3. But for an overall state, we note that

$$|\psi|\psi\rangle = a^2|00\rangle + ab|01\rangle + ab|10\rangle + b^2|11\rangle \tag{9}$$

1.5.2 Blockchain Technology

After the worldwide financial crisis in 2008, Satoshi Nakamoto [1] created an innovative peer-to-peer (P2P) financial system that uses bitcoin as a digital cryptocurrency, taking advantage of recent technological advancements. At first, the idea of central power interfering with economic structure did not excite many people. Blockchain is the name of the technology used to put this system into place [6]. The method that these primitives use to make sure that the records are correct is called automated trust. Additionally, the database of records is publicly accessible and is referred to as a distributed ledger. The blockchain's main constituents are listed below.

- The blockchain uses a distributed network of thousands of devices to store data. Data is therefore very resistant to technical issues and hacking attempts. The dispersed network can assist in preventing single-point failure.
- In a traditional payment system, a bank, credit card company, or payment provider is needed. All nodes in the blockchain must mine to verify transactions. Blockchain is a "trustless system" because of this.
- Once a block is added to the blockchain, not even its owner can change or remove it.
- A transaction is only valid if it encounters the smart contract's requirements; if the condition is satisfied, then the contract is executed automatically.
- Inception. The entire history of debt is always present in a blockchain network.

1.6 Quantum Computing and Security

Classical theory can't explain quantum superposition and entanglement. Feynman said quantum computers can simulate quantum merchant services (QMS) [2]. Traditional Turing machine memory models are insufficient for simulating quantum merchant services (QMS). Classical reorientation becomes quantum superposition. Q-bits power quantum computers. This Q-bit superposes $|0\rangle$ and $|1\rangle$. Q-bits encrypted Entangles Two Q-bits. Entanglement allows measuring one Q-bit while determining another. Any two-level quantum system can contain a Q-bit, like a magnetic field, photon, or spin. Quantum Q-bits aren't 0/1. Superposition allows both states [7]. Four bits make 16 combinations. Q-bits store 16 values. Twenty is 1,000,000-valued. Ensuring data security in the digital age is crucial, especially with the growing use of the Internet.

1.7   An Amalgamation of Quantum and Blockchain Security

Blockchain technology can provide protection to the healthcare industry and offer benefits to medical professionals. For blockchain transactions in healthcare, patient IDs are required, and data is transmitted through APIs. Smart contracts can be used to process patient-related transactions, and the public ID is used for these transactions. Patients can provide their private key to healthcare providers to access their data, which is secured using ordinary encryption until the private key is used. However, with quantum computers, faster algorithms can factorize encryption, which is where quantum encryption comes in to prevent attacks.

Chapter Objectives
- Explore the potential applications of quantum technologies in addressing environmental challenges.
- Investigate the role of blockchain technologies in promoting sustainable development.
- Explore the synergy between quantum computing and decentralized energy systems.
- Delve into the application of blockchain in establishing transparent and ethical supply chains.
- Present case studies and real-world implementations that demonstrate the successful integration of quantum and blockchain technologies in sustainability initiatives.

Chapter Organization
Section 1 commences by establishing the context by emphasizing the imperative nature of inventive resolutions to pressing worldwide sustainability dilemmas. Introductory remarks underscore the urgent requirement for innovative strategies to mitigate environmental hazards. Section 2 delves into an in-depth analysis of blockchain technology and quantum computation. In this article, the transformative capabilities of quantum computing in the context of sustainability initiatives become apparent in regards to waste reduction, energy optimization, and efficiency. Concurrently, we explore the potential of blockchain technology in facilitating the establishment of sustainable and transparent supply chains, placing particular emphasis on the secure implementation of quantum cryptography. The third section presents the research methodology in a systematic fashion. This consists of a thorough analysis of error correction, quantum states, and other pertinent aspects of blockchain technology. In conclusion, Sect. 4 provides a synopsis of the perspectives and ramifications that have been acquired over the course of the chapter. In order to further advance sustainability objectives, Sect. 5 additionally suggests directions for future research and development. In order to maintain transparency and bolster the fundamentals of our investigation, the chapter culminates in an exhaustive citation of sources.

2   Review Literature

2.1   Quantum Mechanics

The greatest precise and comprehensive explanation of the recognized universe is quantum mechanics. Additionally, it serves as the foundation for thinking about quantum multiplication and quantum info. There is no prerequisite for understanding quantum mechanics. Contrary to its reputation as a challenging subject, quantum mechanics is simple to understand [8]. With this grounding, even someone with no prior knowledge of the subject can start solving basic problems in a few hours. Table 1 briefly explains the investment and research trends for quantum computation [9].

2.2   Application Superdense Coding

A straightforward but stunning application of basic quantum mechanics is superdense coding. It expertly integrates the basic ideas of fundamental quantum mechanics discussed in earlier sections, demonstrating how quantum physics can be leveraged for information processing [10]. Superdense coding consists of two parties that are distant from one another and are unadventurously referred to as "A" and "B." Their objective is to communicate some conventional information [3] from Alice to Bob as shown in Fig. 4. This work can infer from superdense coding that the answer to this question is true. Let's say A and B first divide a few qubits in the entangled state:

$$|\psi\rangle = \frac{|00\rangle + |01\rangle}{\sqrt{2}} \tag{10}$$

According to Fig. 4, A initially possesses ownership of the initial qubit, though B is in ownership of the next qubit. Noting that $|\psi\rangle$ is a secure state, it need not be fixed by A sending B any qubits. The entangled state might also be created by a third party by transmitting one of the qubits to A and the other to B ahead of time.

2.3   The Density Operator

Let's move on to the group of quantum operators that are regularly employed. This work substitutes a container enclosing the dispatch for an operative spanning the stripe to represent a single-qubit operator with an operative container straddling two quantum ropes for a binary gate. The unary and binary operators are shown in the following examples.

Table 1 Quantum insider company and university research area defined [9]

Eleven quantum research institutions		
ID	Name	Find out more about quantum
01	IBM	IBM's quantum computer hardware and simulators play a vital role in quantum machine learning to analyze high-energy physics data at the LHC
02	Massachusetts Institute of Technology	MIT researchers were engaged in quantum computing in 2022. Physical review research published reports on quantum computers, room-temperature photonic logical qubits, and Majorana modes from segmented Fermi surfaces
03	Harvard University	Harvard alumni have made contributions to preprint servers like ArXiv, nature methods, and chemical physics in the area of biological sciences at QC lab
04	Max Planck Society	MPS is in charge of quantum computer research this year. On a single superconducting qubit, the researchers have unveiled quantum logic gates with merging quantum network modules with a two-dimensional Cloquet lattice
05	University of Chicago	In 2022, University of Chicago researchers published three articles: 1. "Engineering Dynamical Sweet Spots to Protect Qubits from 1 / f Noise" 2. "Quantum Solver of Contracted Eigenvalue Equations for Scalable Molecular Simulations on Quantum Computing Devices" 3. "Orchestrated Trios: Compiling for Efficient Communication in 3 Qubit Quantum Programs"
06	Chinese Academy of Sciences	Recent quantum computing findings include "observation of energy-resolved many-body localization" and quantum statistical mechanics with VARs and QCs. Quantum computing labs were established
07	University of California, Berkeley	Quantum computer materials for trapped-ion, quantum approximation optimization challenges in nonplanar graph through a superconducting processor
08	University of Maryland, College Park	UoM quantum research aided IEEE's magic: physical review A, microwave superconductivity, and conformal field theories
09	Princeton University	Princeton researchers announced a spin digitizer for cavity-coupled silicon triple quantum dots and a material platform enabling superconducting transmon qubits to obtain the impressive coherence times surpassing 0.3 ms
10	Google Quantum Computer Research Center	In physical review X, quantum, and nature communications, business researchers have published. Using 20 million noisy qubits, 2048-bit RSA numbers can be factored in 8 h
11	University of Tokyo	"Quantum Machine Learning in High-Energy Physics for Event Classification" and "Blueprint for a Scalable Photonic Fault-Tolerant Quantum Computer" were both published by the University of Tokyo

(continued)

Table 1 (continued)

Top 20 quantum investors

ID	Name	Investment	ID	Name	Investment
01	A&E Investments	£179M	11	High-Tech Grunderfonds	€2.3 M
02	Airbus Ventures	$55 M	12	Lux Capital	$4 M
03	Alchemist Accelerator	$3 M	13	Osage University Partners	$55 M
04	Amadeus Capital Partners	£650,000	14	Parkwalk Advisors	£750 K
05	Bloomberg Beta	$225 M	15	Prelude Ventures	$21 M
06	Dcvc (Data Collective)	$40 M	16	Robert Bosch Venture Capital	$21 M
07	Entrepreneur First (Ef)	$2.1 M	17	SGInnovate	$1.9 M
08	Felicis Ventures	$4 M	18	Shasta Ventures	$1.9 M
09	Fenox Venture Capital	$6.5 M	19	Summer Capital	$8.5 M
10	Goldman Sachs	$6.5 M	20	Tim Draper	$32 M

Quantum insider investor details

ID	Investor	Type	Founded	Quantum portfolio	Country
01	180 Degree Capital	Listed vehicle	1981	D-Wave Systems	United States
02	360 Capital	Venture capital	1997	C12 Quantum Electronics	France
03	3i Group	Venture capital	1945	DenseLight	United Kingdom
04	3to1 Capital	Angel investor	2021	BosonQ Psi	India
05	500 Southeast Asia	Venture capital	2014	Atomionics	Singapore
06	500 Startups	Venture capital	2010	Atomionics	United States
07	6th Horizon	Venture capital	2009	Atomionics	
08	7percent Ventures	Investor network	2014	Universal Quantum	United Kingdom
09	808 Ventures	Venture capital	2016	Equal1	Australia
10	8090 Partners	Family office	2020	Luminous Computing	United States

Twelve best quantum computing universities

ID	Name	Working quantum computing
01	The University of Waterloo	The University of Waterloo went all-in on quantum computing, unlike other universities that didn't offer degrees, like cat adoption agencies that turned away the Schrodinger family. Success! 2002 gave Mike Lazaridis. He financed Waterloo's Perimeter Institute for Theoretical Physics, with 296 researchers and 1500 articles. This university commercializes research
02	University of Oxford	Old Oxford's quantum center. 1985: Deutsch describes quantum computer. Universities displayed NMR quantum computers. Qubit guru. Quantum research is praised. Oxford dominates quantum computing theory, technology, and ethics
03	Harvard University Harvard Quantum Initiative	Quantum Computing Institute (HQI) quantum theory is difficult for scientists and engineers. The first "quantum revolution" produced MRI, intercontinental communication, and GPS navigation. HQI members revolt

(continued)

Table 1 (continued)

Twelve best quantum computing universities		
ID	Name	Working quantum computing
04	Mit Center For Theoretical Physics	MIT's research. Affects quantum data. Institution advances QI/QC using theoretical physics. Quantum algorithms, complexity, and QIT are researched at MIT. MIT developed quantum computing
05	National University of Singapore and Nanyang	Singapore's Centre for Quantum Technologies creates quantum devices. Invented quantum computing. Using quantum computing, communications, and sensing
06	University of California Berkeley	Berkeley studies quantum algorithms, cryptography, information theory, control, and quantum devices
07	University of Maryland Joint Quantum Institute	JQI oversees NIST, LPS, and UMD physics quantum research (LPS). Each university conducts quantum computing research
08	University of Science and Technology of China	Focus on quantum optics and information. Quantum computing, superconducting quantum computing, cold-atom quantum simulation, and quantum metrology. Complex research tools
09	University of Chicago and Chicago Quantum Exchange	Chicago Quantum Exchange aids academic and commercial quantum projects (CQE). They funded quantum computer research. CQE emphasizes information sharing, teamwork, and partnerships
10	University of Sydney Australia	The University of Waterloo went all-in on quantum computing, unlike other universities that didn't offer degrees, like cat adoption agencies that turned away the Schrodinger family. Two hundred ninety-six persons and 1500 papers comprise the quantum computing powerhouse. This university commercializes research
11	Quantum Applications and Research	Old Oxford's quantum center. 1985: Deutsch describes quantum computer. Universities displayed NMR quantum computers
12	University of Innsbruck Quantum	We help scientists and engineers uncover quantum applications

Quantum corporations				
ID	Company	Description	Region	Founded
01	1Qbit	1Qbit makes hardware-agnostic quantum software	Americas	2012
02	A Star Quantum	Quantum	APAC	2018
03	ADVA	ADVA provides smart communications and technology. Future generations will benefit from scalable networks with software-automated solutions	Americas	1994
04	AEM	AEM's semiconductor and electronics test solutions use best-in-class technologies, methods, and customer support	APAC	2000
05	AEMtec	AEMtec provides sophisticated microsystems, optoelectronics, and packaging services	EMEA	2000

(continued)

Table 1 (continued)

Quantum corporations

ID	Company	Description	Region	Founded
06	AIXTRON	AIXTRON develops semiconductor deposition equipment, process engineering, consultancy, training, and continuous client support and after-sales support	EMEA	1983
07	AMS Technologies	AMS Technologies provides specialized solutions. Power and thermal management. Quantum produces laser diodes	EMEA	1982
08	AMTE Power	Thurso, Scotland, is AMTE Power's headquarters (previously AGM Batteries Limited). Since 1997, they've made lithium-ion batteries for automotive, aerospace, defense, oil and gas, and energy storage	EMEA	2013
09	AOSense	AOSense makes atom optic equipment for accurate timekeeping, gravity measurement, and navigation	Americas	2004
10	APITech	APITech is a top global technology company that creates and produces RF/microwave, microelectronics, and security solutions	Americas	1999
11	ASML	ASML's headquarters are in the Netherlands. They make lithography equipment used to fabricate computer chips	EMEA	1984
12	AT&T	AT&T offers wireless, local, and long-distance services. AT&T leads research in quantum network technologies and communication	Americas	1093

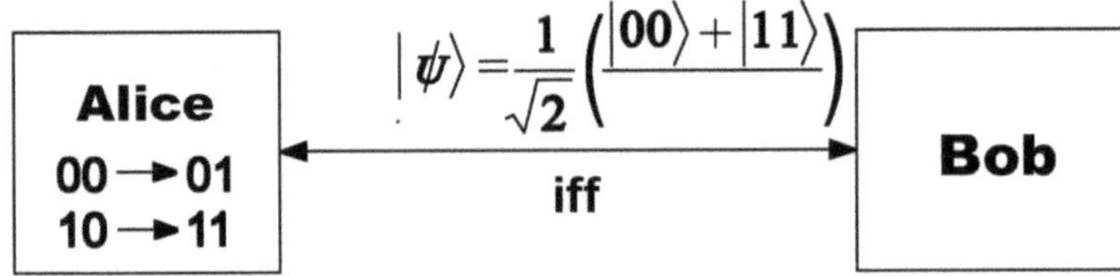

Fig. 4 Superdense coding's A to send B two classical bits

3 Methodology

3.1 Quantum State

Quantum mechanics can be used to name any classical (i.e., non-quantum) computer because it is the base for all corporal production. However, a traditional computer claims that quantum mechanics has enough money to carry out its schemes while not taking advantage of specific possessions.

1. How can the state's structure be described?
2. Describe entanglement.
3. How are reversibility, computation, and physical systems constructed?

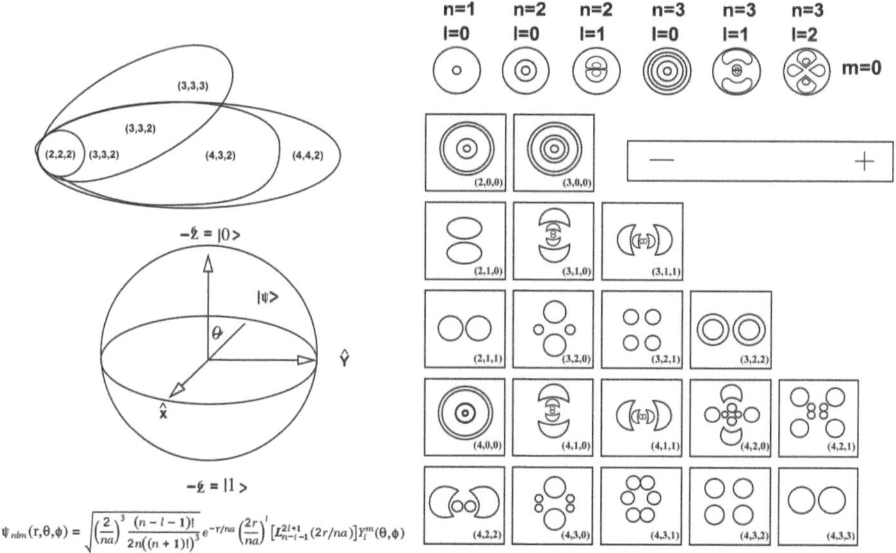

Fig. 5 Force and temperature, built to sense quantum occurrences

Systems are undefined before to an amount, but after we ration them, they are in a definite state. Systems lack a specified state before measurement but assume a specific and unambiguous state once measured. If our system is limited to operating in one of two discrete states, we can represent these conditions in Dirac notation as $|0\rangle$ and $|1\rangle$, as described [11] in Fig. 5. A linear mixture of these states can then be used to characterize a principle of superposition of states, as shown in Fig. 5:

$$\frac{1}{\sqrt{2}}|0\rangle + \frac{1}{\sqrt{2}}|1\rangle \tag{11}$$

Superposition

This work can illustrate the superposition of states using the fundamental feature of the division of light. There is no preferred path for the opposition in practically all of the sunlit we encounter in daily life, such that from the sun. One light opposition, such as vertical opposition, which we might denote as $|\uparrow\rangle$, can be chosen using a single splitting filter. An orthogonal state to vertical opposition is flat opposition, which is represented by the symbol $|\rightarrow\rangle$ [12]. These conditions work as the foundation for any opposition to light. The system's condition is denoted by the Greek letter $|\psi\rangle$:

$$|\psi\rangle = \alpha |\uparrow\rangle + |\rightarrow\rangle \tag{12}$$

The possibility of a state emerging after dimension is the modulus shape of the size of a state in a principle of superposition of states. Additionally, the superposition's rectangles of all conceivable states' breadths added together equal 1. Thus, the state details are as follows $|\psi\rangle = \alpha\,|0\rangle + \beta\,|1\rangle$:

$$\alpha\,|0\rangle^2 + \beta\,|1\rangle^2 = 1 \tag{13}$$

States and Vectors

In classical physics, understanding a system's current state suggests all the information required to predict the system's future. Quantum systems are not entirely predictable, as this work learned in the previous lesson. The meaning of quantum states is distinct from that of classical states. Simply put, a quantum state is perceptive to the extent that its organizational structure can be detected. We discussed creating the state of a spin using a tackle in the previous chapter [13]. Indirectly, we anticipated that there would be no further minute information to posit or quantify regarding the status of the spin.

Whether the unpredictable behavior is caused by incompleteness in what we refer to as a quantum state is the obvious question to raise. There are numerous viewpoints on this subject. Here are several examples:

Representing the Spin States

It's time to use state vectors to try our hand at representing spin states. This work objective is to create a picture that encompasses what we know about the behavior of spins. The procedure will now be more instinctive than correct. Based on our prior research, this work does our best to fit the pieces together. Please take your time reading this section. You'll benefit; I assure you.

Any state might be stated as a linear superposition for the two basis vectors, which could choose at random as $|u\rangle$ and $|d\rangle$, and a superposition takes ideal time. Let's designate a general state with the symbol $|A\rangle$. This can be written as an equation:

$$|A\rangle = \alpha_u\,|u\rangle + \alpha_d\,|d\rangle \tag{14}$$

where αu and αd represent the mechanisms of $|A\rangle$ along $|u\rangle$ and $|d\rangle$, respectively. This work can identify the mechanisms of $|A\rangle$ mathematically as follows:

$$\alpha_u = \langle u|A\rangle,\ \alpha_d = \langle d|A\rangle \tag{15}$$

Along the X-Axis

As previously stated, any spin state can be expressed as a linear mixture of the basic courses $|u|$ and $|d\rangle$. Let's attempt taking ownership of radius and length of qubit the $|r\rangle$ and $|l\rangle$ vectors, which represent spins arranged along the x-axis. Let's begin with $|r\rangle$. If A initially prepares $|r\rangle$ and is then alternated to measure z, as you recall from Lecture 1, there will be equal chances for up and down. Consequently, $\alpha_u{*}\alpha_u$ and $\alpha_d{*}\alpha_d$ must both equal ½. An easy vector that adheres to this principle is:

$$|r\rangle = \frac{1}{\sqrt{2}}|u\rangle + \frac{1}{\sqrt{2}}|d\rangle \qquad (16)$$

There is some uncertainty in this high quality, but, as we shall see in the future, it is less than the uncertainty in our best set of careful x- and y-axis directions.

Along the Y-Axis
Finally, this brings us to the ket quantum state vectors |i⟩ and |o| for the concerned spins along the y-axis. Let's examine the circumstances they must manage. First,

$$\langle i|0\rangle = 0 \qquad (17)$$

According to this condition, up and down are also represented by orthogonal vectors, just as in and out are. According to the laws of physics, if the spin is in, it cannot be out.

Additional restrictions apply to the vectors |i⟩ and |o⟩. With the help of the relationships described by the mathematical outcomes of our experiments, we can write the following:

$$\langle o|u\rangle\langle u|o\rangle = \frac{1}{2}, \langle o|d\rangle\langle d|o\rangle\frac{1}{2}, \langle i|u\rangle\langle u|i\rangle = \frac{1}{2}, \langle i|d\rangle\langle d|i\rangle = \frac{1}{2} \qquad (18)$$

Counting Parameters
The number of self-governing parameters required to characterize a system is a constant concern. For instance, each of the general synchronizes (referred to as q|i⟩) we employed in Volume I characterized a degree of autonomy. This method relieves us of the laborious task of formulating obvious equations to express physical constraints [14]. In a similar vein, our next objective is to determine how many physically distinct states there are for a spin. To demonstrate that you also receive the same answer, I'll do it twice.

As I mentioned previously, we will eventually find that a state vector's physical properties are independent of the full phase factor. This indicates that one of the three permanent parameters is used, leaving just two of the same number of parameters as are required to hypothesize a path in three dimensions. The phrase, hence, contains adequate self-determination:

$$\alpha_u |u\rangle + \alpha_d |d\rangle \qquad (19)$$

Two angles are used to represent the orientation of a point concerning the origin in a sphere-shaped memory. Latitude and longitude provide yet another illustration. However, there are only two possible outcomes of a trial along any axis, to explain all the various positioning of a spin.

Representing the Spin States as Column Vectors
With the help of the intellectual forms of our state vectors, such as |u⟩ and |d⟩ and so on, this work has so far been intelligent enough to study a lot. These ideas allow

us to concentrate on mathematical correlations without being distracted by unnecessary details. However, shortly we will have to perform in-depth spin state calculations; therefore, this work has to write our state vectors in column form. The column symbols are not all the same because of the "phase in consequence"; therefore, we'll try to pick the most modest and appropriate ones we can find [15].

As is customary, this work begins with $|u\rangle$ and $|d\rangle$. Both component length and reciprocal orthogonality are required. These ratios are in the following two columns:

$$|u\rangle = \begin{pmatrix} 0 \\ 1 \end{pmatrix} \qquad (20)$$

$$|d\rangle = \begin{pmatrix} 0 \\ 1 \end{pmatrix} \qquad (21)$$

Putting It All Together

This talk has covered a lot of territory for us. Before continuing, let's evaluate our performance. Our objective was to create knowledge of spins and vector spaces. In the process of figuring out how to use vectors to characterize spin states, we learned what information a state vector contains (and does not contain). Here is a summary of what we accomplished.

Entanglement

If the dimension of one system is more solidly tied to the state of the other system, then associations are standard world. In other words, the two system states cannot be divided [16]. Later in this book, we shall go through the specific exact definitions of separability and entanglement.

Does the amount of energy lost during the execution of a fundamental calculation have a minor bound? We now accept that there is such a boundary, thanks to Landauer and others; this is known as Landauer's assured (LB). The energy charge of removing n bits is more accurately given as nkT ln 2, where ln 2 is, of course, the natural log of 2 (0:69315) [17] and k is the Boltzmann continual. T is the temperature in Kelvin of the heated bowl next to the computing equipment. The minimal quantity of energy required for a permanent multiplication is this limit.

Quantum Gate

A basic quantum circuit employs a few qubits which are known as the quantum logic gate. There are parallels between typical logic gates for predictable digital and quantum computers. Quantum gates are tunable with a direct conversion that can take place between reversible global logic gates like quantum logic gates and Toffoli gate. Further, quantum logic gates are composed of unitary matrices [18].

The vast majority of quantum gates make use of either one or two-qubit regions. It is clear from this that orthonormal matrices of either 2 x 2 or 4 x 4 can be used to describe quantum gates.

SWAP Gate

Two qubits are used in the SWAP gate. The SWAP gate, which is articulated in basis states, switches the states of the two complex qubits:

$$
\text{SWAP} = \begin{pmatrix} 1 & 0 & 0 & 0 \\ 0 & 0 & 1 & 0 \\ 0 & 1 & 0 & 0 \\ 0 & 0 & 0 & 1 \end{pmatrix} \tag{22}
$$

1. Qubits 2.
2. X q [0].
3. H q [1].
4. display # display state previously swaps procedure.
5. SWAP q [0], q [1].
6. display # presentation state after swap procedure.

Hadamard Gate

By mapping the origin states $|0\rangle$ to $\dfrac{|0\rangle + |1\rangle}{\sqrt{2}}$ and $|1\rangle$ to $\dfrac{|0\rangle - |1\rangle}{\sqrt{2}}$, the Hadamard gate creates an equivalent superposition of the two basis states. $H = \dfrac{1}{\sqrt{2}}\begin{pmatrix} 1 & 1 \\ 1 & -1 \end{pmatrix}$.

1. H q [0] # perform Hadamard gate on qubit 0.
2. H q [1:2,5] # perform Hadamard gate on qubits 1,2 and 5.

Decompositions

1. Another express 180° turns around the x-axis, and then a 90° turns around the y-axis. So, $H = X Y^{1/2}$.
2. Provided are helpful XY-decompositions (also envisioned below).

   (a) $H = X Y^{1/2}$.
   (b) $H = Y^{-1/2} X$.

3. Useful YZ-decompositions are.

   (a) $H = Z Y^{-1/2}$.
   (b) $H = Y^{1/2} Z$.

CNOT Gate

CNOT gate two-qubit procedure, with the initial qubit often designated as the controller qubit and the next qubit as the board qubit. Further, the CNOT gate is expressed as follows:

1. Once the controller qubit is in state $|1\rangle$, shrubberies the control qubit immobile and completes a Pauli-X gate on the goal qubit.
2. It does not alter the target qubit when the qubit state is unchanged $|0\rangle$.

$$CNOT = \begin{pmatrix} 1 & 0 & 0 & 0 \\ 0 & 1 & 0 & 0 \\ 0 & 0 & 0 & 1 \\ 0 & 0 & 1 & 0 \end{pmatrix} \tag{23}$$

1. Qubits 2.
2. H q [0] # Create a superposition state by performing a Hadamard gate on qubit 0.
3. Using a CNOT gate q [0], q [1] # entangle equal numbers of qubits.

Quantum Algorithm

What kind of computations is possible with quantum circuits? In what way does that period compare to the computations that container be done with conventional logical circuits? Is there a commission that a quantum computer could be able to complete more effectively than a classical computer? This unit addresses these issues, describes in what way to run classical calculations on quantum computers, provides some instances of difficulties where quantum computers outperform classical computers, and provides a quick overview of the existing quantum algorithms [19].

Berstein-Vazirani Algorithm

The Bernstein-Vazirani quantum algorithm finds the hidden bit string in a black box function. The issue is as follows: given a black box function as $f(x)$, where x is an n-bit string and $f(x) = a \cdot x + b$ (where an is an n-bit string and b is a single bit), calculate the values of a and b. The classical algorithm involves n black box function queries to find a and b. The Bernstein-Vazirani algorithm solves the problem with one black box function query.

The Algorithm

- Initialize a single-qubit and n-qubit quantum register.
- Apply a Hadamard gate to each qubit in the n-qubit register to superpose all n-bit strings.
- A Hadamard gate creates a | − | state in the single-qubit register.
- Black box both registers.
- Apply a Hadamard gate to each n-qubit register qubit again.
- Calculate a form the n-qubit register.

Given n inputs and an unknown function,

$$f\left\{\left(x_{n1}, x_{n2}, \ldots\ldots, x_1, x_0\right)\right\} \tag{24}$$

Mysterious, non-negative, less-than 2^n number. Allow $f(X)$ to accept slightly additional such integer X and the sum of X times a, modulo 2. The function's output is as follows:

$$a.x = a_0 x_0 \oplus a_1 x_1 \oplus a_2 x_2 \ldots\ldots \tag{25}$$

Shor's Algorithm

Shor's quantum algorithm solves integer factorization, a difficult classical computer problem. The renowned quantum algorithm was discovered by Peter Shor in 1994. Given a positive integer N, find its prime factors. Shor's method operates in polynomial time on a quantum computer, while traditional algorithms take exponential time.

Shor's approach exploits the fact that quantum computers solve the period-finding issue exponentially quicker than classical computers. Quantum computers use the QFT to efficiently find the period of the function f(x). Shor's algorithm is the first quantum algorithm that solves problems quicker than classical algorithms. Many current cryptographic protocols depend on the difficulty of factoring huge integers. Shor's algorithm could break these protocols, prompting post-quantum cryptography research.

RSA cryptography: Let's say A wants to transmit a confidential message over the Internet. A spiteful watcher, Eve, may quite possibly interfere with a message's transmission along the way. If A is transferring a note, this is difficult but innocent; but, if A is transmitting her credit card number, there is a problem. How are messages sent securely over the Internet?

The study of secret code production and resistance is known as cryptography. Writing secret codes is referred to be cryptography, whereas defying those codes is referred to as cryptography. The secure transmission of evidence over the Internet is made possible by the widely used RSA cryptography. Rivest, Shah, and Adelman were three of the early proponents of this development [20].

The period of a function: In some ways, the challenge of determining the equivalent to the period of a function is tricky factorizing the production of two huge prime integers. Consider raising a number, such as 2, to ever-advanced powers, and before attracting the effect modulo the product of two prime integers, such as $91 = 13 \cdot 7$, to get a concept of what the history of a function would be. For instance, we have

2^0	2^0 (mod 91)	1	2^4	2^4 (mod 91)	16
2^1	2^1 (mod 91)	2	2^5	2^5 (mod 91)	32
2^2	2^2 (mod 91)	4	2^6	2^6 (mod 91)	64
2^3	2^3 (mod 91)	8	2^7	2^7 (mod 91)	37

We can see that the modulo operation prevents the numbers from producing continuously. The highest power, for instance, is 2^7 (mod 91/ = 37 when 2^6 (mod 91/ = 64).

Grover's Algorithm

The hypothetical speed limit for classical algorithms is quadratically quicker than Grover's approach to discovering a component in an unordered set. The needed element defines a function, and if that element is present, the function is judged to be true for that element. The next step is to locate all fundamentals that pass the test with the least amount of oracle requests possible [21].

The data assembly contains N elements, each of which has n = log N bits. To begin, we convert $|0\rangle^{\otimes n}$ using a Hadamard method to get the superposition state:

$$|\psi\rangle = \frac{1}{\sqrt{n}} \sum_{x=0}^{n-1} |x\rangle \tag{26}$$

The Grover operative's application of its mechanisms is efficient. Each of the two Hadamard converts needs n processing steps. The restricted stage change requires O(n) gates and is a measured unitary operation. Depending on the application, the oracle calls can be challenging, but only one call is required for each iteration. The algorithm charts the important steps one more time [4].

- Quantum superposed all database states. Apply a Hadamard gate to n qubits initialized to $|0|$ to create an equal superposition of all conceivable n-bit strings. This produces a state $|s = (1/2^n) * $ n-bit string $|x$.
- Oracle: special unitary transformation. The oracle, U-f, flips solution phase. U-f thus maps $|x|$ to $(-1)^{f(x)}|x$, where f(x) is 1 if x is a solution and 0 otherwise.
- Another unitary transformation D, the diffusion operator, amplifies the state $|s|$ and diminishes the states orthogonal to it. First, each qubit receives a Hadamard gate, then the mean amplitude is flipped, and finally the gates are applied again. $D = 2|s| - I$ is the diffusion operator.
- Repeat steps 2 and 3 approximately O(sqrt(N)) times.
- Measure the qubits to get an n-bit string. The likelihood of measuring a solution state is related to its final state amplitude squared.
- Apply the oracle again to verify the n-bit string's solution. It is likely correct if it returns 1.

Quantum Risk to the Ethereum Blockchain: A Bump in the Road or a Brick Wall?

The utilization of cryptocurrency is growing. They'll have a market cap of more than \$2 trillion by 2021, and they'll be integrated into the established financial system. Cryptography-protected decentralized trust is the foundation of cryptocurrencies rather than a centralized authority. The security of cryptocurrencies is at risk from future quantum computers, which are predicted to break some of the common cryptographic methods for which there are no obvious alternatives. Concerning our past research on Bitcoin, how can the quantum threat to the Ethereum blockchain be measured? What difficulties do corrective measures have?

The Main Focus of This Chapter Will Be to Answer the Following Questions

- What is the impact of quantum computing on cryptocurrencies?
- How do storage and transit attacks differ?
- When is this going to become a problem?

Practical Work

1. The lithium-sulfur (Li-S) battery has the potential to represent a major advancement in battery technology and is currently being developed and optimized by Mercedes-Benz R&D North America. The successful implementation of Li-S

batteries could open up a billion-dollar market. However, extensive chemical testing and engineering expertise are necessary to move from the conceptual design stage to producing Li-S batteries on a large scale.

2. Prior to physical prototyping, it is highly desirable to conduct computer simulations to explore and analyze the diverse array of molecular properties and behaviors. However, the number of quantum states and electron interactions involved in even seemingly simple chemicals like caffeine, which is made up of carbon, hydrogen, nitrogen, and oxygen, can be enormous.

3. The memory and processing capabilities of even the most powerful conventional computers, including the current state-of-the-art supercomputers, can be completely overwhelmed by the creation of a single caffeine simulation. The difficulty of this process only increases as the size and complexity of the molecule and its surroundings grow.

3.2 *Quantum Error Correction*

Quantum computers of the modern era do not have sufficient qubits to do full quantum error correction (QEC), although expertise in QEC is gradually being developed to guide QC and other applications. Duplicating states across a large number of classical bits makes it possible for classical computing to acknowledge and rectify frank errors [22]. As a result of the no-cloning principle, quantum computers are unable to apply this particular direct method.

3.3 *Blindness*

During the transaction, the traditional message M2 is not transferred from A to Ali in its original form. Instead, it is transformed into a quantum sequence A1, A2..., An, along with its associated measurement results, which are then encrypted to construct EKAC(M2) and recorded on the blockchain. Trent possesses the particle Ti and is aware of the states of two more particles, but since TI and T1 are equal, he is unaware of the necessary measurement basis. Trent uses a random basis of $|0\rangle$ or $|1\rangle$ and $|+\rangle$ or $|-\rangle$, or the proxy signer, to obtain the correct answer with a probability of $1/n^2$, according to his calculations. The one-time pad and controlled quantum teleportation ensure that Trent, like B1, is not aware of the message M2. Therefore, M2 is blind to everyone except A and Ali. The protocol has the property of being blind, so to speak.

3.4   Unforgeability

First, we determine that B1's signature cannot be forged by the dishonest insider. Suppose Charlie tries to counterfeit the signature because he is dishonest. Ali's fraud will be seen as a result of the blockchain's features. The keys are all dispersed via quantum key distribution (QKD) protocols as well, preventing others from forgetting B1's signature without the associated key. Similar to Ali, every participant in this technique will experience identical circumstances. As a result, internal attackers are unable to alter the signature. Second, we talk about the forgery that the outside attacker Eve created. Let's say Eve is a spy or an attacker who wants to fake B1's signature. Similarly, to this, Eve needs the key K_{B1C} to forge B1's signature. Eve is unable to fake B's signature as a result. Finally, we use H-gates to assure security throughout the transmission of M1 [23].

3.5   Undeniability

This work integrates the blockchain into our system so that none of the participants may contest the records kept there. To decide where decoy photons should be introduced, we also compute the key's hash value during the transmission of the sequence M1. Additionally, the key is unique and distributed via quantum key distribution (QKD). However, the key is dependent on both the M1 sequence's encoding principles and the associated H-gate procedures. You cannot, in other words, refuse transmission or other processing.

3.6   Known Attacks

The entangle-and-measure attack, as well as the intercept-and-resend and CNOT assaults, is unsuccessful against our protocol. First, we will use an intercept-and-resend attack. Even if Eve is an outside attacker, he is able to intercept Alice's quantum sequence H1KTA; nonetheless, he is unable to locate the decoy photons without the secret key and the hash value. Each decoy photon that Eve measures will result in an error rate of 50% on his part. At the very least one and a half times, Eve will be located. The symbol t is used to denote dummy photons. In the event that t is sufficiently large, the eavesdropping detection system will promptly identify Eve.

3.7 Message Attack

Eve is able to successfully intercept a blockchain transaction that is being sent from A to B through a communication channel; however, she is only able to access the transaction data if she is able to obtain the encrypted text B by possessing all initial parameters, extract the hash code hash 'A which is N28, and correctly determine the number of nodes in the circle N. A and B conduct their transactions involving N, 0, 1, and 2 in a secure location or via tamperproof quantum channels. The precision of the calculations performed by digital devices places a limit of $10(-16)$ on the key space used in the generation of the hash value and the key stream, which increases the security of the communication. As a result, our cybersecurity protocol is resistant to message attacks and is an appropriate choice for blockchain-based smart water utilities.

3.8 No-Message Attack

Let's say Eve sends B a phony blockchain transaction to make him believe A authenticated it. When B receives the transaction, he is unable to quickly determine whether it was initiated by A or Eve. To tackle this problem, B first generates hash code hash$'_A$ from S and cipher-text B using the initial key parameters that he and A exchanged. He next acquires hash$'_B$ for the received plain text and verifies it with hash$'_A$ as per our blockchain framework's verification phase. Finally, the discovery of Eve's existence causes this deal to be abandoned. As a result, our suggested cybersecurity protocol is protected against no-message attacks and may be applied to blockchain-based smart water utilities.

3.9 Man-in-the-Middle Attack

In order to contact and obtain transaction data during this assault, Eve will pretend to be either A or B. When Eve impersonates A, she is unable to use the default parameters of the QHF to conduct a fraudulent transaction with B. This transaction will be authenticated by B, who will then cancel it. Since Eve is unable to discover the vital settings of the QHF, she is unable to violate the transaction data even if she poses as B and communicates with A. As was previously mentioned, the expansive key space of QHF secures transaction data. Even if Eve were to acquire the cipher text and the hash value, she would still be unable to access the transaction data since the whole key parameters are protected by a quantum channel. Attempts at impersonation are thwarted by our cybersecurity solution, which safeguards blockchain-based smart water utilities.

4 Conclusion

In addition to storing and manipulating states that, if represented classically, would correspond to enormous amounts of information, a quantum computer is also capable of doing particular computations exponentially quicker than any classical counterpart. Blockchain is a distributed database that utilizes cryptography to protect it from unauthorized changes. Blockchain platforms use digital signatures that quantum computers can abuse, despite its promise. A party using quantum computation has a little edge in mining rewards due to cryptographic hash algorithms used to build new blocks. Computational biology may undergo a paradigm shift as a result of the possibility for even small quantum computers to outperform the finest supercomputers on some tasks, promising to make both tough and impossible issues regular. Within the next 10 years, practical early quantum processors are anticipated to appear. Therefore, every computational scientist should prioritize learning what quantum computers are capable of. An experimental blockchain architecture that is safe from quantum attacks is presented here. This design makes use of the dispersion of quantum keys across an urban fiber network in order to provide information-theoretically secure authentication. These conclusions answer significant concerns about the viability and scalability of quantum-safe blockchains for usage in commerce and government. This work has made an effort to provide an overview of the technologies that could soon change computational biology in this article. Even though we were unable to cover every subject, this work did present the fundamentals of quantum information processing and talked about three crucial ones.

5 Future Work

- Quantum computing is a rapidly evolving field with significant potential for future applications.
- As technology advances, we are aware that hardware that increases the likelihood of running quantum programs is essential to quantum computing. We need to scale up quantum computers since they may require hundreds of thousands or millions of high-quality qubits. The 433-qubit "Osprey" and 1121-qubit "Condor" processors, anticipated for deployment in 2022 and 2023, will advance single-chip processors and the laws controlling large-scale quantum systems connected to IBM Quantum System Two. We don't want to understand chip-sized quantum computers. Instead, we're creating physics-defying modular processing designs. Table 2 briefly explains the investment and research trends for quantum computation from 2007 to 2021 [17].
- The latest generation of AI is as self-aware as a paper clip. The odd assertion made by a Google engineer that his company's artificial intelligence system "came to life" and the prediction made by Elon Musk, CEO of Tesla, that computers will approach human intelligence by 2029 demonstrate that technology

Table 2 Quantum insider funding [17]

Companies	Primary Classification	Secondary Classification	Date	Founded	FTEs	Country	City	Transaction Type	Reign	Total $	Citation
Scantinel photonics	Premium access	Premium access	2022-11-25 12:19:12	2019	11–50	Germany	UIm	Series A	EMEA	10,000,000	[24]
Xanadu	Premium access	Premium access	2022-07-06 2:43:39	2016	11–50	Canada	Toronto	Series C	Americas	100,000,000	[25]
Infleqtion	Premium access	Premium access	2022-11-30 1:14:6	2007	51–100	United States	Boulder	Series B	Americas	110,000,000	[26]
OTI Lumionics	Premium access	Premium access	2022-07-06 2:43:39	2011	11–50	Canada	Toronto	Series B	Americas	55,000,000	[27]
Qunnect	Premium access	Premium access	2022-07-06 2:43:39	2017	1–10	United States	Stony brook	Series A	Americas	8,000,000	[28]
Quantum computing incorporated	Premium access	Premium access	2022-08-25 1:41:9	2018	None	United States	Leesburg	Other	Americas	7,500,000	[29]
Classiq	Premium access	Premium access	2022-11-28 9:4:30	2018	51–100	Israel	Tel Aviv-Yafo	Series B	EMEA	48,500,000	[30]
Kipu quantum	Premium access	Premium access	2022-07-10 6:17:12	2021	1–10	Germany	Berlin	Other	EMEA	3,000,000	[31]
EeroQ	Premium access	Premium access	2022-08-18 8:57:48	2016	1–10	United States	New York	Seed	Americas	7,250,000	[32]
CamGraPhIC	Premium access	Premium access	2022-09-26 9:47:33	2018	11–50	United Kingdom	Cambridge	Other	EMEA	1,524,600	[33]

still struggles with routine jobs. This includes operating a motor vehicle, particularly in unforeseen circumstances that call for even a small amount of knowledge or comprehension.

- The world's disillusionment in AI experts, or rather, the basic technologies concealed as such, should serve as a sobering reminder. Quantum computing, a more potent emerging technology, has the potential to cause chaos, especially if it is combined with AI. Before it's too late, we urgently need to comprehend the potential effects of this technology, regulate it, and prevent it from falling into the wrong hands. The world must avoid making the same mistakes it did by rejecting to normalize AI.

- QEC is a fast-expanding field of research with ramifications for both QC and the outside world. Classical computation allows for upfront mistake correction by replicating a state across several classical bits. We are unable to use this straightforward method on a quantum computer, however, due to the no-cloning hypothesis in quantum physics [34].

- IBM Quantum is the pioneer in quantum computing. Even the most powerful traditional supercomputers of today and tomorrow may not be able to and will never be able to address some of the crucial problems that this cutting-edge technology is capable of. More than 210 Fortune 500 businesses, fictitious organizations, national labs, and start-ups make up our community of clients and partners. We aim to transform the most intriguing issues into well-received opportunities with the largest convoy of 20+ of the most potent quantum systems in the world [20].

References

1. Whitfield, J. D., Yang, J., Wang, W., Heath, J. T., & Harrison, B. (2022). *Quantum computing 2022* (pp. 1–13) [Online]. Available: http://arxiv.org/abs/2201.09877
2. Nielsen, M. A., & Chuang, I. L. (2012). *Quantum computation and quantum information.* CUP. https://doi.org/10.1017/cbo9780511976667
3. Vogel, V. (2011). *Quantum computation and quantum information,* by M.A. Nielsen and I.L. Chuang, vol. 52, no. 6. https://doi.org/10.1080/00107514.2011.587535.
4. Krantz, P., Kjaergaard, M., Yan, F., Orlando, T. P., Gustavsson, S., & Oliver, W. D. (2019). A quantum engineer's guide to superconducting qubits. *Applied Physics Reviews, 6*(2), 1–67. https://doi.org/10.1063/1.5089550
5. Liu, Q., et al. (2017). Extensible 3D architecture for superconducting quantum computing. *Applied Physics Letters, 110*(23), 1–5. https://doi.org/10.1063/1.4985435
6. Bhavin, M., Tanwar, S., Sharma, N., Tyagi, S., & Kumar, N. (2021). Blockchain and quantum blind signature-based hybrid scheme for healthcare 5.0 applications. *Journal of Information Security and Applications, 56*(December). https://doi.org/10.1016/j.jisa.2020.102673
7. Alzoubi, Y. I., Gill, A., & Mishra, A. (2022). A systematic review of the purposes of Blockchain and fog computing integration: Classification and open issues. *Journal of Cloud Computing, 11*(1). https://doi.org/10.1186/s13677-022-00353-y
8. Rieffel, E., & Polak, W. (2000). An introduction to quantum computing for non-physicists. *ACM Computing Surveys, 32*(3), 300–335. https://doi.org/10.1145/367701.367709

9. All quantum technology data points the quantum inside. Quantum Companies & Research. (2022). https://app.thequantuminsider.com/companies

10. Manzalini, A., & Amoretti, M. (2022). End-to-end entanglement generation strategies: Capacity bounds and impact on quantum key distribution. *Quantum Reports, 4*(3), 251–263. https://doi.org/10.3390/quantum4030017

11. Rinaldi, E., et al. (2022). Matrix-model simulations using quantum computing, deep learning, and Lattice Monte Carlo. *PRX Quantum, 3*(1), 1. https://doi.org/10.1103/PRXQuantum.3.010324

12. Sigov, A., Ratkin, L., & Ivanov, L. A. (2022). Quantum information technology. *Journal of Industrial Information Integration, 28*(May), 100365. https://doi.org/10.1016/j.jii.2022.100365

13. Corsiglia, G., Schermerhorn, B. P., Sadaghiani, H., Villaseñor, A., Pollock, S., & Passante, G. (2022). Exploring student ideas on change of basis in quantum mechanics. *Physical Review Physics Education Research, 18*(1), 10144. https://doi.org/10.1103/PhysRevPhysEducRes.18.010144

14. Panwar, E., Kukunuri, A. N. J., Singh, D., Sharma, A. K., & Kumar, H. (2022). An efficient machine learning enabled non-destructive technique for remote monitoring of sugarcane crop health. *IEEE Access, 10*(June), 75956–75970. https://doi.org/10.1109/ACCESS.2022.3190716

15. Yu, Q., et al. (2022). Feasibility study of quantum computing using trapped electrons. *Physical Review A, 105*(2), 1–11. https://doi.org/10.1103/PhysRevA.105.022420

16. Orús, R., Mugel, S., & Lizaso, E. (2018). Quantum computing for finance: Overview and prospects. *Reviews in Physics, 4*(September), 2019. https://doi.org/10.1016/j.revip.2019.100028

17. Jaeger, L. (2019). *The second quantum revolution: From entanglement to quantum computing and other super-technologies.* Copernicus. https://doi.org/10.1007/978-3-319-98824-5

18. Nagayama, S. (2017). Distributed quantum computing utilizing multiple codes on imperfect hardware. [Online]. Available: http://arxiv.org/abs/1704.02620

19. Lu, S. C., Zheng, Y., Wang, X. T., & Wu, R. B. (2017). Book #10 - Quantum machine learning by P. Wittek (vol. 34, (11).

20. Silva, V. (2018). *Practical quantum computing for developers: Programming quantum rigs in the cloud using Python, quantum assembly language and IBM Qexperience.* Apress. https://doi.org/10.1007/978-1-4842-4218-6

21. Mavroeidis, V., Vishi, K., Zych, M. D., & Jøsang, A. (2018). The impact of quantum computing on present cryptography. *International Journal of Advanced Computer Science and Applications, 9*(3), 405–414. https://doi.org/10.14569/IJACSA.2018.090354

22. Wu, Y., Kolkowitz, S., Puri, S., & Thompson, J. D. (2022). Erasure conversion for fault-tolerant quantum computing in alkaline earth Rydberg atom arrays. *Nature Communications, 13*(1), 1–16. https://doi.org/10.1038/s41467-022-32094-6

23. Lin Gou, X., Shi, R. H., Gao, W., & Wu, M. (2021). A novel quantum E-payment protocol based on blockchain. *Quantum Information Processing, 20*(5), 1–17. https://doi.org/10.1007/s11128-021-03126-9

24. Insider. (2019). Scantinel photonics. https://app.thequantuminsider.com/company/c01ff66f-0646-4e44-8c3c-fcc8a46446af/profile. Accessed 25 Nov2022.

25. Insider. (2016). Xanadu. https://app.thequantuminsider.com/company/579e022a-cfcd-4b34-ae15-92187875e1c2/profile. Accessed 6 Jul 2022.

26. Insider. (2007). Infleqtion. https://app.thequantuminsider.com/company/59e56dd2-d846-46b9-980d-16aae0b1bcc8/profile. Accessed 30 Nov 2022.

27. Insider. (2011). OTI lumionics. https://app.thequantuminsider.com/company/9dfc564b-4d8d-4460-a6e9-8f6f0b48897b/profile. Accessed 6 Jul 2022.

28. Insider. (2017). Qunnect. https://app.thequantuminsider.com/company/b5d89a40-613c-4eef-a657-263cdc00bb37/profile. Accessed 6 Jul 2022.

29. Insider. (2018). Quantum computing incorporated. https://app.thequantuminsider.com/company/b5d89a40-613c-4eef-a657-263cdc00bb37/profile. Accessed 25 Aug 2022.

30. Insider. (2018). Classiq. https://app.thequantuminsider.com/company/b5d89a40-613c-4eef-a657-263cdc00bb37/profile. Accessed 28 Nov 2022.

31. Insider. (2021). Kipu quantum. https://app.thequantuminsider.com/company/ b5d89a40-613c-4eef-a657-263cdc00bb37/profile. Accessed 10 Jul 2022.
32. Insider. (2016). EeroQ. https://app.thequantuminsider.com/company/b5d89a40-613c-4eef- a657-263cdc00bb37/profile. Accessed 18 Aug 2022.
33. Insider. (2018). CamGraPhIC. https://app.thequantuminsider.com/company/ b5d89a40-613c-4eef-a657-263cdc00bb37/profile. Accessed 26 Sep 2022.
34. Steane, A. (1998). Quantum computing. *Reports on Progress in Physics, 61*(2), 117–173. https://doi.org/10.1088/0034-4885/61/2/002

Combating Counterfeit Drugs
in Pharmaceutical Supply Chain (PSC)
Using Hyperledger Fabric Blockchain

Anitha Premkumar and Rajesh Natarajan

Abstract Counterfeit drugs are fake-labeled medicines or substandard medicines. Drugs without active pharmaceutical ingredients (APIs) or those with wrong proposition of APIs are also called counterfeit drugs. These drugs are very harmful to human life. According to the World Health Organization (WHO) report of 2017, 10% of pharmaceutical medicines in India are fake or substandard. Counterfeit drugs find their way into pharmaceutical supply chain (PSC) easily due to security issues and traceability issues in the current system. The software solutions presently in use hold the drug detail in databases on cloud central server and are accessed by active actors such as manufacturers, distributors, and retailers in the supply chain via the Internet. A malicious participant in the supplychain can tamper with the database and introduce the counterfeit drugs at any point. To eliminate this security flaw, a secure system with distributed decentralized digital ledger (DDDL) technology can be implemented. The information in such a system is stored on a digital ledger and shared with all the authorized nodes in the peer-to-peer (p2p) network without worrying about central authority to govern. DDDL is a new form of immutable shared database that can be implemented using blockchain. In DDDL, only authorized actors in the network can perform read/write operations. Such a blockchain-based solution leads to improved security and better trust and transparency.

Here, we have developed a secured system which is built upon the most popular blockchain platform called Hyperledger fabric blockchain (HFB) to identify and eliminate the counterfeit drugs from PSC and thereby avoid risk to the end consumers who need the drugs. A chaincode is a core element of Hyperledger fabric that defines a set of rules and policies for business logic to execute. We have included

A. Premkumar (✉)
Department of Information and Communication Technology, Manipal Institute of Technology Bengaluru, Manipal Academy of Higher Education, Manipal, India, Bangalore, India

R. Natarajan
Information Technology Department, University of Technology and Applied Sciences-Shinas, Shinas, Sultanate of Oman
e-mail: rajesh.natarajan@shct.edu.om

S. Pulipeti et al. (eds.), *Quantum and Blockchain-based Next Generation Sustainable Computing*, Contributions to Environmental Sciences & Innovative Business Technology, https://doi.org/10.1007/978-3-031-58068-0_4

Hyperledger Explorer as benchmark tool for analyzing the performance and efficiency of the system that we designed.

Keywords Pharmaceutical supply chain (PSC) · Counterfeit drug · P2P network · Distributed decentralized digital ledger (DDDL) · Hyperledger fabric blockchain (HFB)

1 Introduction

"Counterfeiting difficulties are happening in numerous ways," as stated in [1]. It can range from selling drugs without the active component to more harmful actions like labeling drugs incorrectly. A counterfeit drug, as defined by the WHO study [2], is one that has been purposefully and unlawfully mislabeled with regard to its identity and/or source. One of the biggest issues affecting both wealthy and developing nations is the counterfeiting of drugs. According to WHO, the range of counterfeit pharmaceuticals is believed to be between 30% and 70% in developing nations and around 7% in industrialized countries. Africa has a serious issue with fake medications. According to a research report, it is typical for counterfeit medications to pass through 30 different hands before they are consumed [3]. India and China are the top 2 countries where fake medications are produced and sent around the world, according to a USTR (United States Trade Representative) report [4]. The actors involved in the distribution and transportation of produced pharmaceuticals are currently not fully disclosed by the supply chain process. Lack of data transparency in the supply chain increases the likelihood that fake pharmaceuticals may enter the pharmaceutical supply chain (PSC) [5] and complicates the drug recall procedure. Table 1 discusses the various crime reports based on incidences of pharma crime that occurred in India between 2012 and 2020.

To stop these unlawful scenarios from happening, a reliable system that provides PSC security, traceability, and transparency as well as easy access to drug information at any time and place is required. This type of reliable system can be created with the use of cutting-edge IT technology known as the blockchain [6–9], which is well-liked among corporate applications in the banking, healthcare, supply chain, education, and government sectors. The data on blockchain is kept more safe via a cryptographic method. The data is encrypted and authenticated using two keys via the cryptographic technique [10]. There are two ways to implement blockchain technology: public blockchain and private blockchain. By generating their own public address, everyone can participate in the system network via public blockchain [11]. Every node in the network must take part in the consensus process in order to join the public blockchain [12]. Bitcoin [13, 14] and Ethereum [15] are popular public blockchain systems. Private blockchain is used within an organization; to carry out the digital transactions, a private network is set up and members from the same organization are added. The process of adding members will be done

Table 1 Pharmaceutical crime incidents in India between 2012 and 2020

Incident	Place & year	Incident Details	Crime Report	Pharmaceutical Crime Evidence
Sub-Standard Drugs were identified	Jan 2020, India	Drugs failed to meet the quality test because of presence of foreign matter	1336 Samples tested 1286 were declared of standard quality, 49 samples including eye drops, painkillers even antibiotics as not of standard	Regulator flags 49 drug samples as 'not of standard quality'
Unauthorized sale of Drugs	Dec 2019, India.	Sale of drugs without checking prescription	Death of two students who allegedly overdosed on prescription painkillers	Medical stores inspected
Spurious drugs Seized	Jul 2016, Luknow, India	Fake medicines were sold in the market	Seized spurious drugs worth of Rs 5 Lakhs	Spurious drugs worth Rs 5 lakh seized from Aminabad market
Substandard medicines were identified	May 2012, India	No active ingredients , re-labeled and sold	Drug samples were tested and found that out of 48082 samples, 2186 samples failed the quality test and were found to be spurious or adulterated	1 in 3 dodgy drugs from Maha, TN

by membership service provider of the organization. Hyperledger [16–19] and R3Corda [20] are the examples of private blockchain implementations.

The research article is divided into various sections:

- Section 2 analyzes the various blockchain-based healthcare systems
- Section 3 discusses about design methodology
- Section 4 elaborates the implementation details and working model of chaincode.
- Section 5 discusses about the proposed system's result.
- Section 6 visualizes and monitors the operations of network parameters of a proposed system using Hyperledger Explorer.
- Section 7 discusses about the paper conclusion

2 Background Work

In this section, the focus is to cover blockchain-based healthcare system and pharmaceutical supply chain. The authors of [21] presented the issues and challenges related to PSC. According to the study, pharma companies are subjected to various risks. These risks affect the quality of drugs and quality of supply chain. It affects the delivery of drugs to the customer at the correct time and place.

In [22], the author discussed about the widespread of fake drugs in India and challenges faced by pharmacovigilance. In this study, the author explains about counterfeit drug definitions given by the WHO and Drug and Cosmetic Act, USA. Also in this paper, the author mentions incidents that have occurred in legal trade from India where counterfeit drugs were identified on different occasions.

In this paper [23], the author deals with online counterfeiting and explains how the Internet is helping counterfeiters by providing a huge customer base at low risk. It discusses about threats that are posed by counterfeit drugs to public safety. Included in the paper are dangers of SSFFC medicines, actions, and the way forward.

In [24], the author explains the global scenario of drug counterfeiting with examples of pharmaceutical crime incidents and analyzes the unique problems of drug counterfeiting in India as compared with other source countries. The author gives some suggestions and recommendations to overcome drug counterfeiting which are (1) implementing RFID tags on each medicine and (2) tracking and tracing the mechanism of medicine.

In [25], the author highlights some of the pharma crime incidents which took place in various part of India between 2002 and 2004. And the author discussed about SFFC (substandard/falsified/fake/counterfeit) drugs, and how it creates health hazards to the public was also mentioned in this paper.

According to Gupta et al. [26], counterfeit drugs are of various types which include fake products, incorrect measure of the active ingredients, misbranded medications, drugs not containing active ingredients, presence of harmful substances, and containing unlisted active ingredients. Also, the author highlighted certain statistical data about counterfeit drugs. He mentions various technologies to prevent counterfeit drugs which include holograms, RFID, 2D barcode, and QR codes.

The Indian Drugs Authentication under the government of India has launched the project called DAVA [27] 2017, which aims to address the drug counterfeit problem by designing integrated platform for track and trace of drugs. This project was designed to use serialization for all pharmaceutical products. This system requires the manufacturers to follow three packaging levels: (a) primary package which is an initial level of packaging that it has direct contact with the product; (b) secondary package which is a level of packaging that contains one or more primary packages grouped together in a single item; and (c) tertiary package which is a level of packaging also called logistical unit that is shipped. Manufacturers must aggregate this packaging information as parent-child information and enter into DAVA database. One of the operational challenges of the system is that it leads to data integrity issues due to the construction of DAVA database. In [43], the author has mentioned the challenges of supply chain management such as (1) multiple and disconnected supply chain actors, (2) costly data reconciliation process, and (3) lack of product traceability. Many leading enterprises are looking at adopting blockchain-based supply chain management that enables collaboration between suppliers, customers, and business partners. In paper [28], the author discusses about the applications of blockchain, various challenges faced in using blockchain in real-life applications, and future perspectives of the same. Mainly the author discussed about applications of blockchain in healthcare domain. Various healthcare solutions based on blockchain are discussed here—security threat mitigation using blockchain, managing storage capacity, interoperability issues, standardization, and social challenges. The author provides a SWOT analysis for blockchain in healthcare. For drug supply chain management system, the novel medical blockchain model was proposed for smart hospitals using Hyperledger fabric [29]. Hyperledger fabric handles secure drug supply chain records. A smart contract, the heart of Hyperledger fabric is designed using solidity programming language to provide security, transparency, and privacy. This permits only valid access to drug supply chain records and patient electronic health records. The author has proposed a front-end web application to cooperate with blockchain network and provides services to the front end.

The proposed solution for PSC [30] will be built using a private blockchain. The drug details are entered into the system by authorized members/nodes of an organization. In PSC, manufacturer, distributor, retailer, and transporter are the authorized and active entities. They are liable for correctness and completeness of drug data stored in PSC database.

The proposed solution approach for PSC using blockchain proposes to use Hyperledger fabric [16–19] with chaincode. There are three different phases of consensus in Hyperledger fabric:

- Endorsement—process of verifying data upon submission.
- Ordering—upon acceptance through endorsement, commit the order to ledger.
- Validation—validate the correctness of the blocks.

Smart contract in Hyperledger fabric is called chaincode [31]. A chaincode handles the enterprise logic created and agreed by the members of the organization.

3   System Design

3.1   Ecosystem

Blockchain is the most trustworthy technology for healthcare systems due to its decentralized nature. In the ecosystem of blockchain-based PSC, a drug information is stored in a digital ledger by various actors like manufacturers, distributors, retailers, and transporters. In this system, digital ledger is kept as separate database. This benefits all the peers who are part of the organization; it enables drug verification, drug identification, and elimination of fake drugs in PSC at every level. Each peer will have certain access permissions to the database based on their credentials. These permissions are set and implemented by Hyperledger's chaincode which ensures integrity of data and its privacy. The system allows the authorized nodes only to access the decentralized digital ledger and perform authorized transactions. Data consistency is achieved by implementing consensus algorithm on each node of the system. This algorithm helps to achieve correctness and consistency of data stored at digital ledger.

The ecosystem of blockchain enabled PSC process is illustrated in Fig. 1. The process flow can be described through the following steps.

Step 1: List of medicines are manufactured by manufacturer, and medicine information is entered into decentralized distributed ledger.
Step 2: Post manufacturer data entry, medicines are now ready to be sent to distributor's locations.
Step 3: Distributor verifies all drug information received from manufacturer by performing read operation in the decentralized ledger.

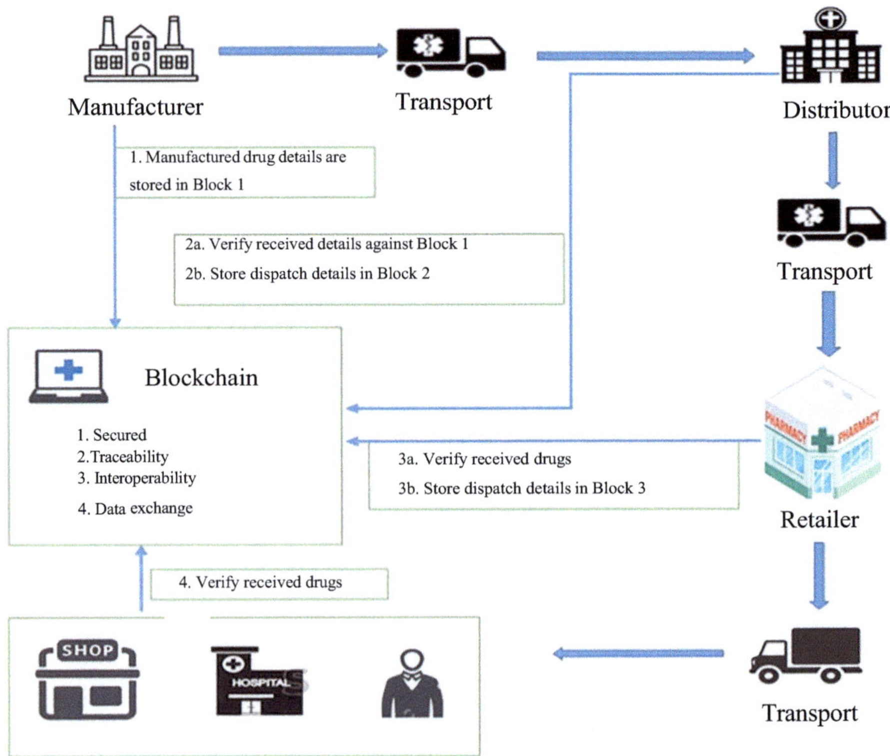

Fig. 1 Ecosystem of blockchain enabled pharma supply chain

Step 4: After verification is done, the distributor enters the received drug details into digital ledger, through write operation. Post entry, a drug is ready for shipping to next point.

Step 5: Retailer collects and verifies the drug details; the origin and the delivery route of the drug can be verified by performing read operation to the decentralized ledger. Post verification, the retailer makes an entry to the ledger about the received drug details.

Step 6: Transporter plays a vital role, shifting the drug to various points. Every time the shipment moves from point to point, transporter verifies and makes an entry into the distributed decentralized ledger.

3.2 System Design

The most significant feature of the solution is the ability to record the transaction information on immutable, decentralized ledger. The transaction details are hashed and encrypted to ensure the security. The proposed blockchain-based PSC design is shown in Fig. 2.

The proposed solution includes two components, i.e., client SDK and blockchain network. A client SDK receives a request from client for transaction and submits it to the blockchain network. The blockchain network includes a smart contract, network nodes, and digital ledger. A request from client is converted into query by client application. This query will invoke the smart contract for reading and writing the transactions into the decentralized ledger after getting the confirmation from all the participating nodes. Once the confirmation is received from network nodes, smart contract updates

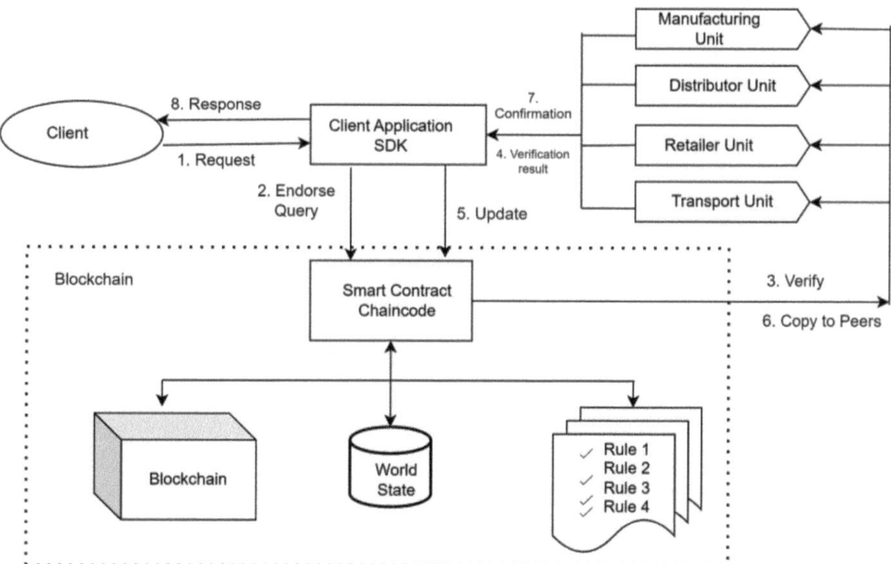

Fig. 2 System design of proposed blockchain-based PSC

the blockchain ledger. Post this, smart contract broadcasts the copy of updated ledger to the network participants. Finally, the client receives a response from blockchain network via client SDK. The nodes in the blockchain network fall into three categories which are orderer peer, endorser peer, and ordinary peer. A node that performs read/write action into the ledger via smart contract for reading and updating the transaction information is called an orderer peer. The orderer peer can update the ledger only after getting the approval from all the other peers in the network. All the confirmations are given by endorser peers. There are two database structures used here for storing the transaction details: the blockchain and world-state database. A blockchain database stores a complete log of transactions initiated by network participants, i.e., manufacturer, distributor, retailer, and transporter. A world-state database is used to store current transactions. A given network node can verify the drug transaction status at any given point of time. To ensure the security and privacy, the proposed system creates single private channel in which manufacturer, distributor, retailer, and transporter are connected. The main idea of setting up the private channel is to distribute the transaction data with the channel nodes directly without revealing to other nodes which are from outside the blockchain network. Thus, having a private channel within system network improves the overall security. The system is designed based on API service calls that use smart contract chaincode and digital ledger as middleware. In our proposed model, the transactions are initiated by manufacturers, distributors, retailers, and transporters through API service calls. The system is built on permissioned private network which makes it different from other such blockchain-based solutions. The feature of permissioned private network allows only authorized nodes to participate and register through the network using membership service provider (MSP). This service provides node enrollment, node validation, and digital signature generation for individual nodes who are participating in the blockchain network.

3.3 Smart Contract Chaincode

Chaincode, one of the essential parts of the blockchain network, is in charge of maintaining the transaction data's sequential sequence of receipt. It authenticates each application for a transaction before approval and saves the results as data blocks in the blockchain network. For instance, the manufacturer node uses chaincode to enter the drug's name, serial number, manufacturing date, and expiration date into the digital ledger. A chaincode enables the client to read just pertinent data from the digital ledger, such as the drug's name and serial number. Based on the data that the node is looking for, a specific chaincode function is called. By using chaincode functions, every legitimate node can submit a request to the digital ledger. A chaincode responds by carrying out the transactions and updating the ledger. Figure 3 shows how client SDK-based chaincode allows network users to communicate with the ledger. JavaScript is used to create the chaincode. The programming language used for designing smart contract is relatively new one. Therefore, the learning process is very hard. The overall performance of transaction execution is limited by smart contract due to the programming language used.

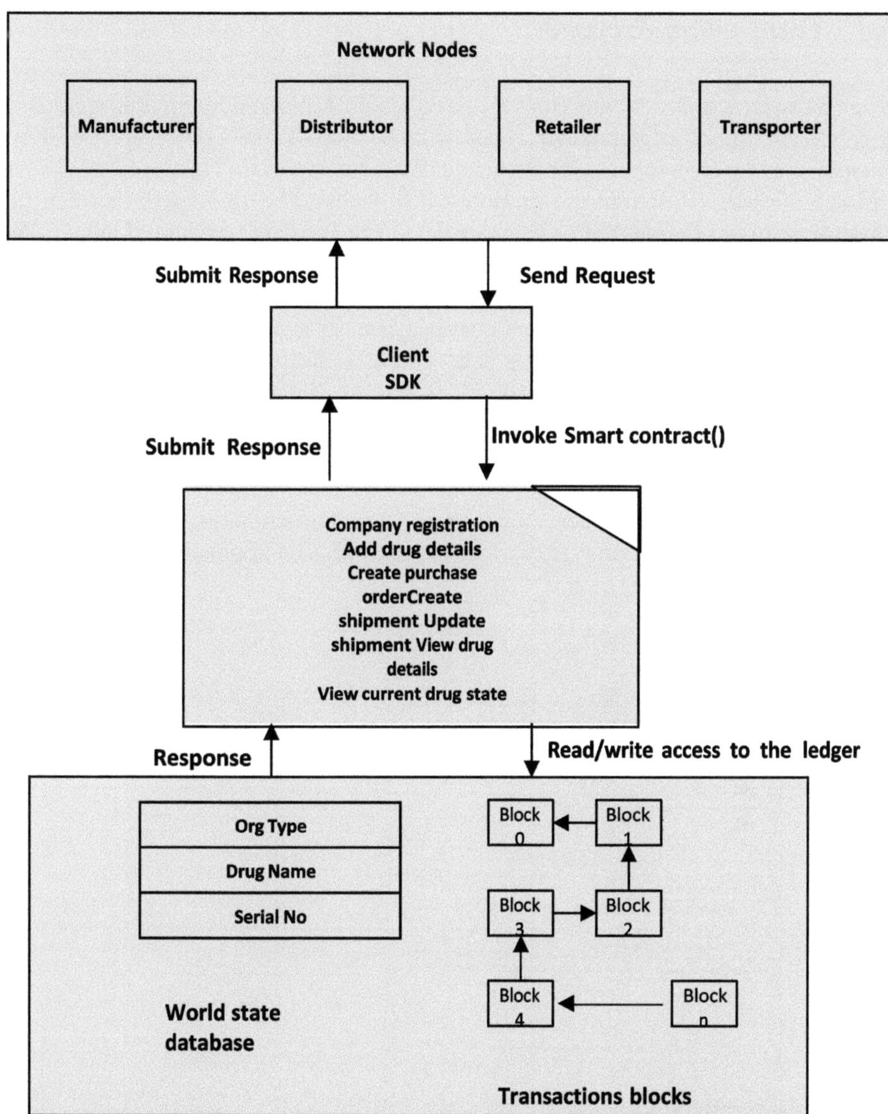

Fig. 3 Network participants interact with ledger using smart contract chaincode

3.4   Transaction Proposal

Every valid data transaction is converted into tamperproof and immutable records of data into the digital ledger on every node with the help of chaincode. Figure 4 shows the role of every individual component and how communication takes place between various components within the proposed PSC network with respect to when the pharmacist/orderer peer verifies the drug details and updates the digital ledger. The client application verifies the user credentials for retailer node to authorize the sending data transaction proposals. Each participating peer in the network can either be endorser or orderer. The position of endorser peer is to verify the transaction proposal and authorize it. The orderer peer validates the results in responses to the transaction and writes the transaction block into the ledger. The endorser peer can act as a committer node who receives and executes the transaction proposal by invoking a smart contract chaincode without updating the ledger. The endorser peer performs a read operation from world-state database during the execution of transaction proposal. The peer then signs the transaction proposal and sends response to client application for further processing. The application SDK collects

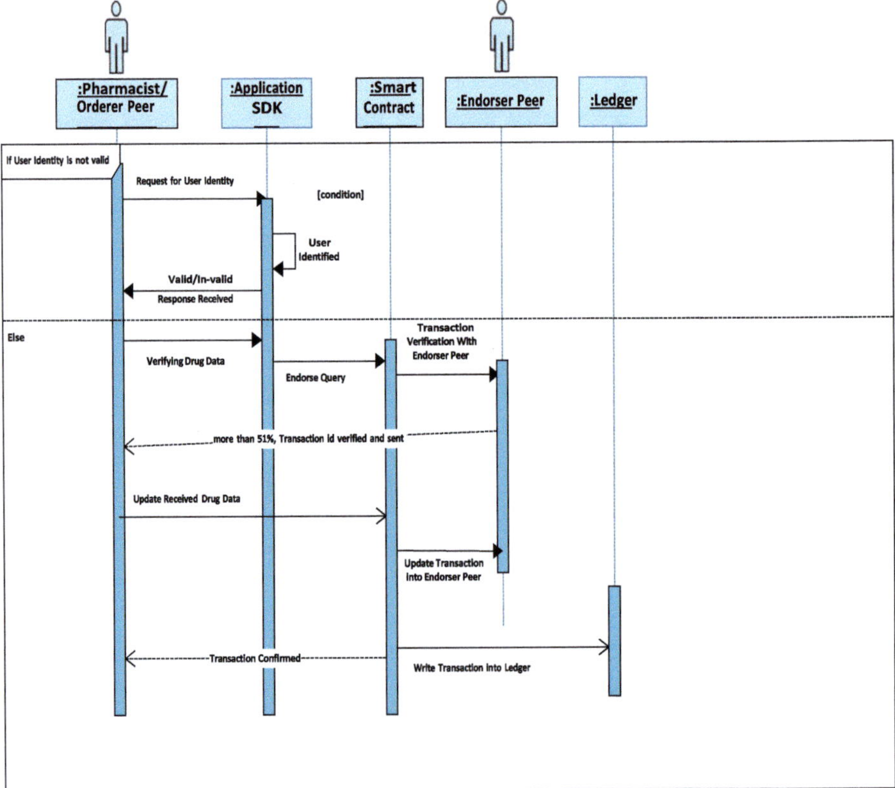

Fig. 4 Illustration of possible drug verification process by pharmacist

the signed transaction and hands it over to smart contract chaincode. The smart contract chaincode sends it to the committer peer or orderer peer in order to write the valid transactions as blocks into the ledger. Now the endorser peer will update the world-state database.

4 Implementation Approach

4.1 Environment Setup on Amazon Web Services (AWS)

Setting up the blockchain network on a dedicated infrastructure will be very expensive and time-consuming. Hence, we propose to use AWS-based platform and thereby reduce complexity, save time, and leverage sophisticated cloud features in the proposed PSC. AWS [42] is a cloud-based offering from Amazon which provides various services to create and deploy applications in a public cloud environment. It provides unmatched scalability and sophistication. AWS has a wide array of on-demand services to end customers on a pay-as-you-go pricing model. The AWS has provided the managed IT services for our proposed work. These managed IT service includes EC2 (elastic cloud computing) which can be used for a variety of workloads. EC2 provides virtual computing environments which have the balance of services such as compute, memory, and networking resources for our proposed system. Table 2 lists the AWS on-demand infrastructure used in the solution.

A common tool to access EC2 instance from AWS is Secure Shell (SSH). SSH connection needs two components, namely, SSH client and SSH server. SSH client is installed on local machine to connect to remote server. The client uses credentials to initiate the secure connection with remote server. If the credentials are valid, SSH server will establish a secured connection with client to AWS server as shown in Fig. 5.

4.2 Execution Procedures

Once the network connection is established with AWS, the prerequisite software can be installed on a local host to support the hyperledger fabric execution. The list of prerequisites is mentioned in Table 3.

Table 2 Resources from AWS for implementing for a proposed system

Hardware services	Description
Operating system	Ubuntu Linux 18.4.1 LTS
Memory	8/16 GB
CPU	Intel core i5
Type of instance	t2. Medium

                                                    A. Premkumar and R. Natarajan

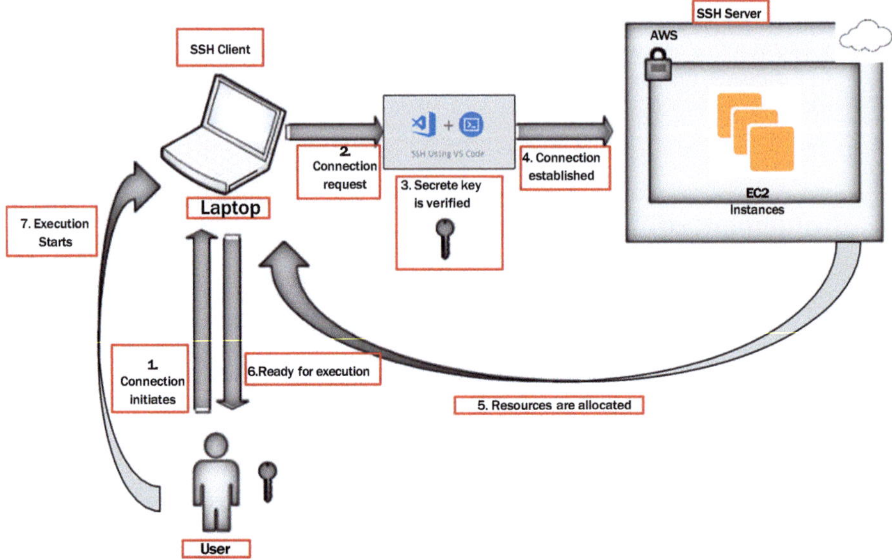

Fig. 5 Accessing AWS infrastructure using SSH protocol

Table 3 List of prerequisite software required for our system

List of prerequisite software	Purpose	Linux commands
Docker 19.03.4	It contains necessary libraries and dependencies for running an application	#sudo apt install docker.io
Docker Compose 1.21.0	For running multi-container applications. It has YAML file	#sudo apt install docker.compose
Python 2.7	High level language	#sudo apt install python
Curl 7.65.3	It supports data from/to a server using Internet protocols	#sudo apt install curl
CLI	A tool to interact with fabric network	
npm 6.12.1	To install third-party libraries	#sudo apt install npm
node 12.13.1	A platform to build fabric network applications	#sudo apt install nodejs
Golang	For writing Hyperledger fabric applications	#sudo apt install golang.go

4.3 PSC Blockchain: Component View

The participants, namely, manufacturer, distributor, retailer, and transporter, can invoke the transaction queries to update the drug information on the ledger with help of a smart contract. The list of participants and transaction queries of the proposed PSC network is shown in Fig. 6. Hyperledger fabric framework helps to create a network with these participants and allowed each one to do permissioned read

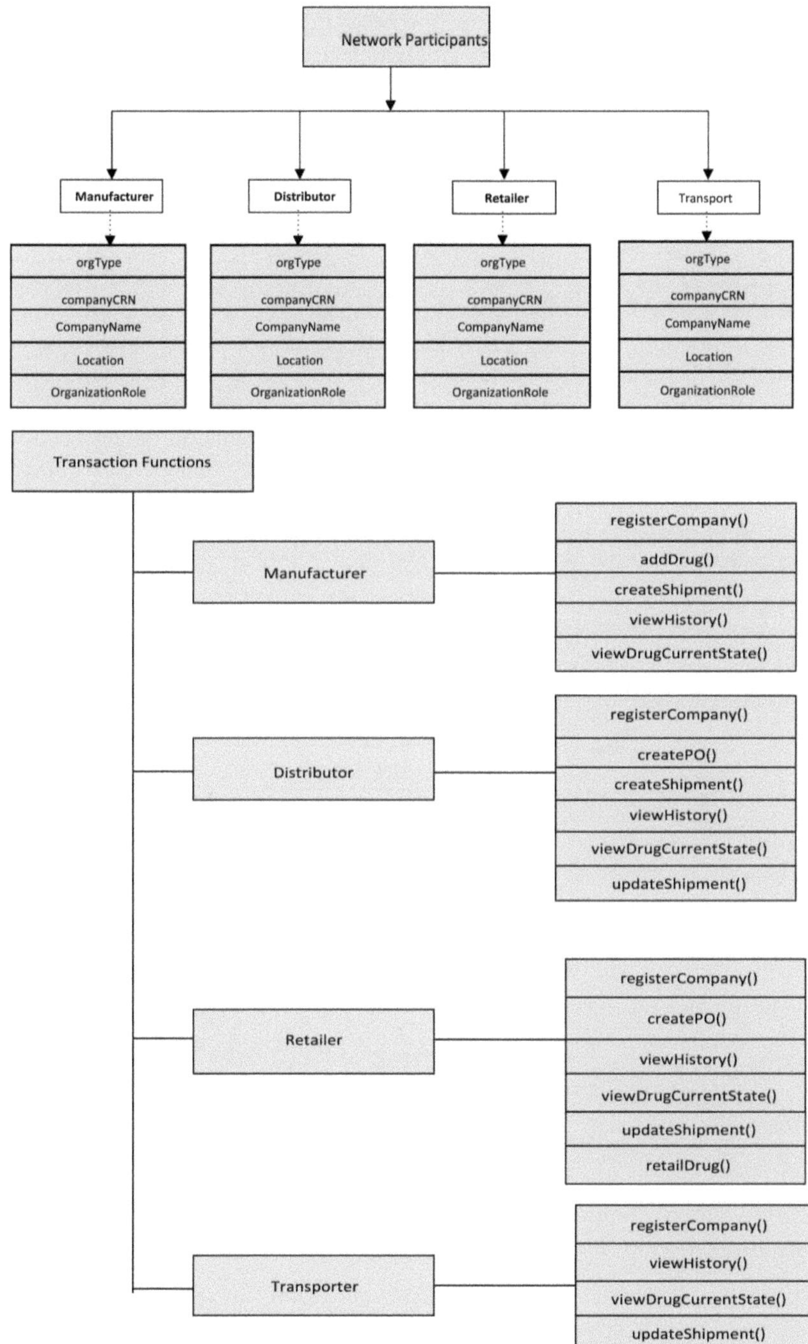

Fig. 6 Network components

and write access to the digital ledger. The network model creates a private channel with these participants/organization, and two peers per organization are included in it for ease of use. The private channel keeps the data more secured and provides transparency to the participants in the network.

4.4 Network Definitions and Transaction Query Definitions

Transactions are performed by network participants in order to verify the drug status, drug ownership, and drug provenance at any given time. Hyperledger fabric maintains all the transactions as encrypted data to achieve the network security. Smart contract chaincode helps to keep the encrypted transaction data in digital ledger and provides restricted access to network participants. In the proposed solution, network participants are manufacturer, distributor, retailer, and transporter. Every participant owns drug asset and can submit transactions. Tables 4 and 5 show network definitions and transaction definitions used in proposed blockchain enabled PSC.

Table 4 Network definition used in proposed blockchain enabled PSC

Network variable	Type
orgType	char
companyCRN	char
companyName	char
location	char
organizationRole	char
drugName	char
mfgDate	date
expDate	date
buyerCRN	char
sellerCRN	char
quantity	int
listOfAssets	char
transporterCRN	char
retailerCRN	char
customerName	char
customerAadhar	char
serialNo	int

Table 5 List of query definitions in blockchain enabled PSC

List of queries	Category	Use	Actors
registerComapny()	Transaction	Register the company	Manufacturer Distributor Retailer Transporter
addDrug()	Transaction	Add the drugs	Manufacturer
createPO()	Transaction	Create a purchase order	Distributor Retailer
createShipment()	Transaction	Create a drug shipment	Manufacturer Distributor
updateShipment()	Transaction	Update the shipment	Transporter
retailDrug()	Transaction	Retail the drug	Retailer
viewHistory()	Transaction	History of the drug	Manufacturer Distributor Retailer Transporter
viewDrugCurrentState()	Transaction	Current state of the drug	Manufacturer Distributor Retailer Transporter

4.5   *Decentralized Distributed Digital Ledger Structure*

Decentralized distributed ledger contains two components as mentioned previously: blockchain and a WS (world-state) database. The WS database holds the current value of ledger state due to which node can access the current value, i.e., the current status of drug information rather than going through the entire transaction log details. This is very flexible that every time the state changes, it updates the state value automatically. There are two world-state database implementations available based on the data format. The LevelDB is the default state database which stores data in key-value format. This database is located inside the node of a network and embedded inside the operating system. The alternative database is CouchDB. CouchDB stores the data in JSON format. Figure 7 shows the drug as a ledger state. Drug is stored in CouchDB. This stores data in both key-value pair and JSON format.

5   Implementation Results

Screenshots of the execution results are provided in this section. The execution of the solution requires both an execution environment and the prerequisite software listed in Tables 3 and 4. Chaincode needs to be deployed on the blockchain network, which has a number of parties, including a manufacturer, distributor, retailer, and

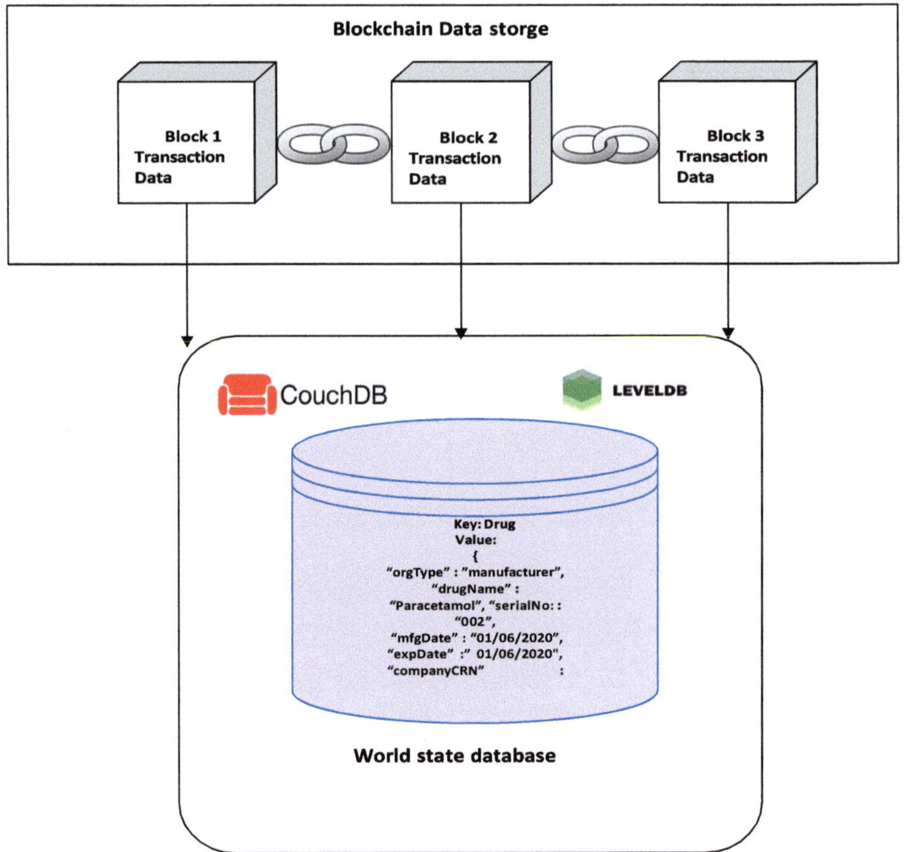

Fig. 7 Structure of decentralized distributer digital ledger of the blockchain enabled PSC model

transporter, in order to carry out transactions. To handle the network strain and traffic, we have introduced two peers for each company in this approach. The membership service provider gives each organization member of the network a unique identity so they may be recognized within the blockchain network. Figure 8 shows how the distributor organization is given MSP so they may identify themselves in the network. MSP can also be made available to all organizations.

The chaincode for registering organizations onto the blockchain network is called registerCompany (), and it is supposed to be invoked with the following information: orgType, companyCRN, companyName, location, and organization Role. Postman and the command-line interface are used to launch this query. Figure 9 shows that the manufacturer organization is listed as having registered with companyCRN. In the event that an organization attempts to register using a CRN that has already been registered, chaincode modifies the system and displays the message "Invalid COMPANY detail- Another firm with this CRN already exists."

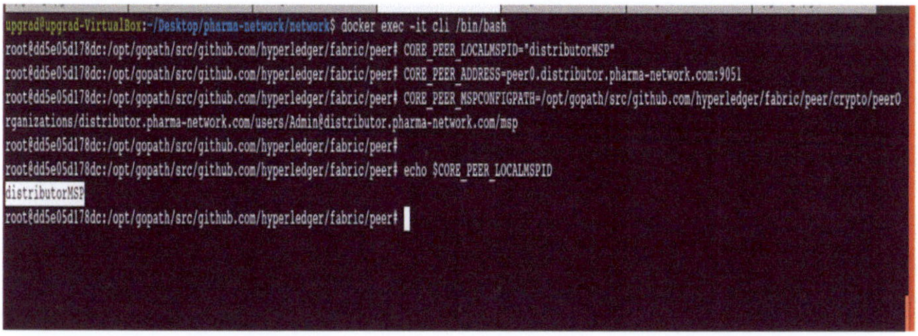

Fig. 8 Distributor MSP process

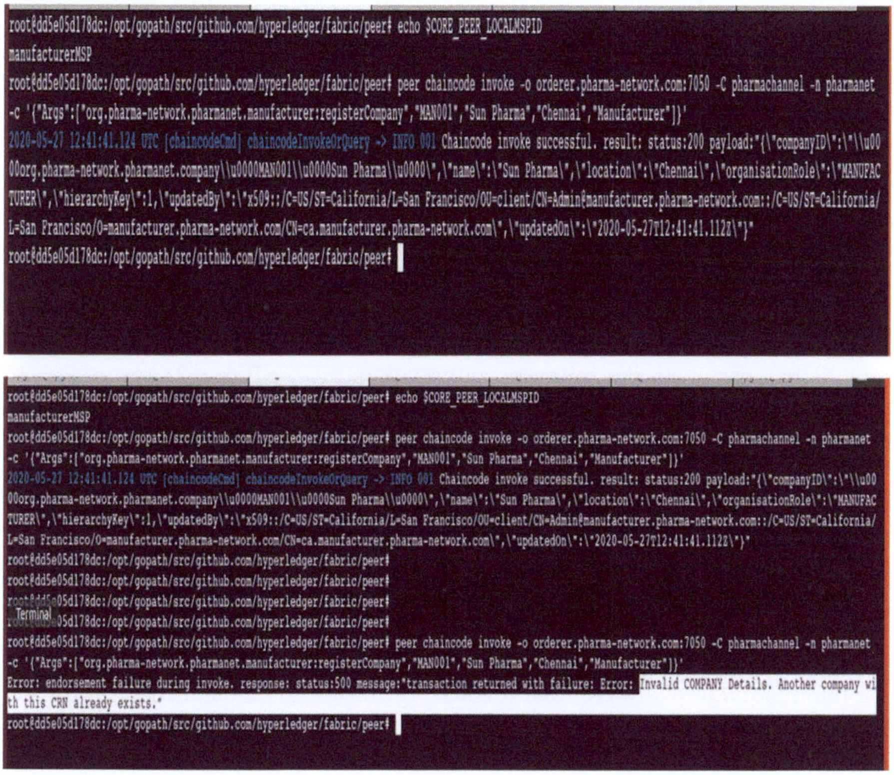

Fig. 9 A registerCompany() query invoking the chaincode through CLI

This demonstrates that our suggested paradigm prevents the system from storing duplicate data.

Another query named addDrug() invokes the chaincode to add the drug details into the ledger. This query is initiated by manufacturer organization. This addDrug() function includes the following drug details such as orgType, drugName, serialNo, mfgDate, expDate, and companyCRN. In Fig. 10, it shows that manufacturerMSP is invoking chaincode via addDrug() query to add the drug information into the ledger. Here we have done validations on various scenarios such as (1) drug already exists, (2) invalid initiator, (3) invalid CRN, and (4) invalid organization role to validate our proposed model that it does not allow any incorrect data to be stored at the network. This feature of chaincode allows our proposed system to verify every transaction data which gets stored in digital ledger. Thus, the system can eliminate incorrect drug data and makes sure that all data stored at ledger is proper. Hence, our system achieves data authenticity.

Another query called createPO() invokes chaincode for creating purchase order for the specified drugs. This query is initiated by distributor and retailer organizations. The createPO() includes orgType, buyerName, sellerCRN, drugName, and quantity. The distributorMSP and retailerMSP invoke the chaincode using createPO() query for the purchase order, and it is shown that the transaction was successful. In Fig. 11, various scenarios are executed to validate our proposed model are shown. These scenarios are as follows: (1) BUYER can't be Manufacturer; (2) invalid BUYER CRN; (3) No DRUG available with specified name TEST; and (4) SELLER does not own this DRUG. This shows that the proposed system does not allow any network participants to perform invalid transactions into ledger. Thus, system keeps only valid data as blocks into the ledger which can be shared by other valid participants in the network. Hence, proposed system achieves data integrity.

A query called createShipment() is initiated by manufacturer and distributor to transport the assets to various locations. This query invokes the chaincode to perform the task of shipment. In Fig. 12, it shows that the manufacturer and distributor initiate the query with the following information: (1) orgType, (2) buyerCRN, (3) drugName, (4) listOfAssets, and (5) transporterCRN. And system validations are done with respect to different scenarios such as (1) invalid initiator, (2) invalid BUYER CRN, (3) invalid transporter CRN, and (4) invalid operation. No PO is raised for specified items. For all the scenario validations, chaincode transactions were returned with failure. This means that our system does not allow any invalid transaction to be stored in the network database.

Another query is called updateShipment() which updates the shipment details with the following information: (1) orgType, (2) buyerCRN, (3) drugName, and (4) transporterCRN. This is initiated by transporter organization. This query updates the information to distributor and retailer organizations via chaincode. In Fig. 13, transporter updates the shipment details into the ledger via chaincode. Different validations are done on proposed system such as (a) invalid initiator, (b) invalid BUYER CRN, (c) invalid transporter CRN, and (d) can't find any SHIPMENT with specified BUYER and DRUG.

Another query is called viewHistory() which invokes the chaincode to view the full history of drug asset by all organizations that are present in the network. This query includes the following information: (1) orgType, (2) drugName, and (3) serialNo. In Fig. 14, it shows the manufacturer's viewHistory.

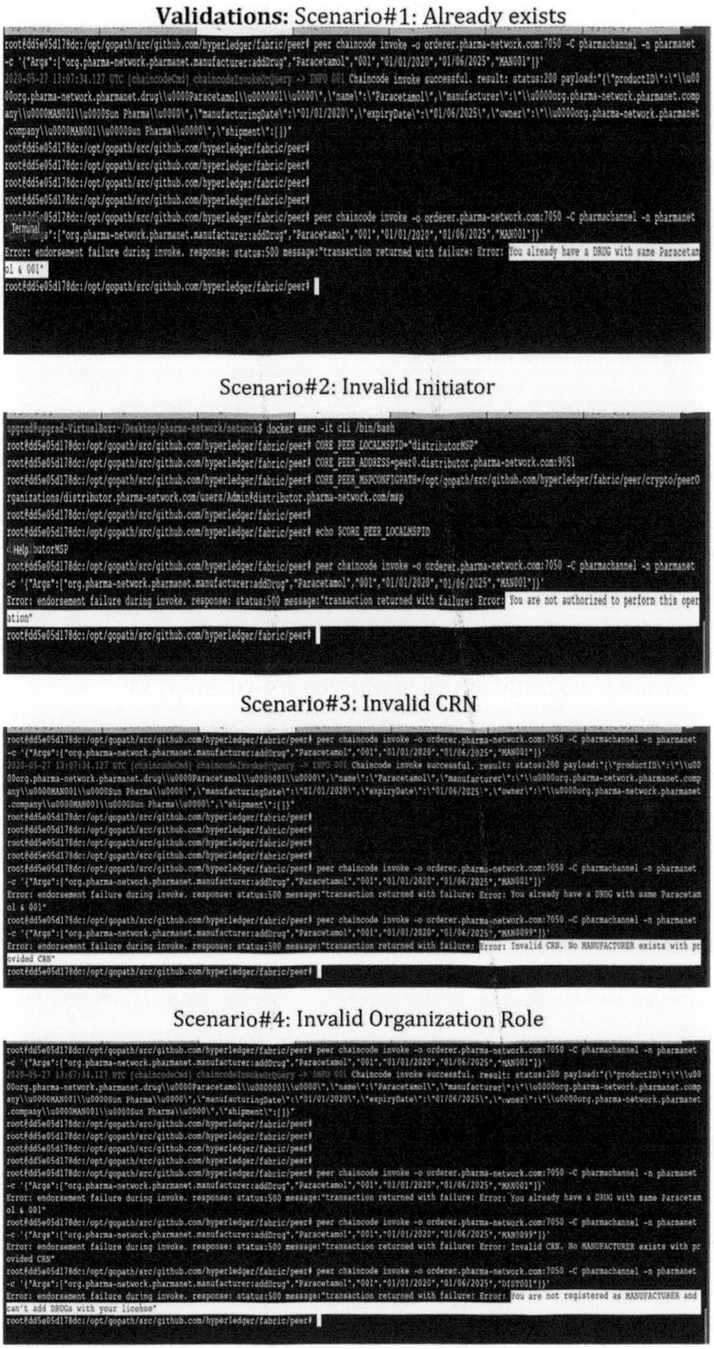

Validations: Scenario#1: Already exists

Scenario#2: Invalid Initiator

Scenario#3: Invalid CRN

Scenario#4: Invalid Organization Role

Fig. 10 An addDrug() query invoking chaincode through CLI

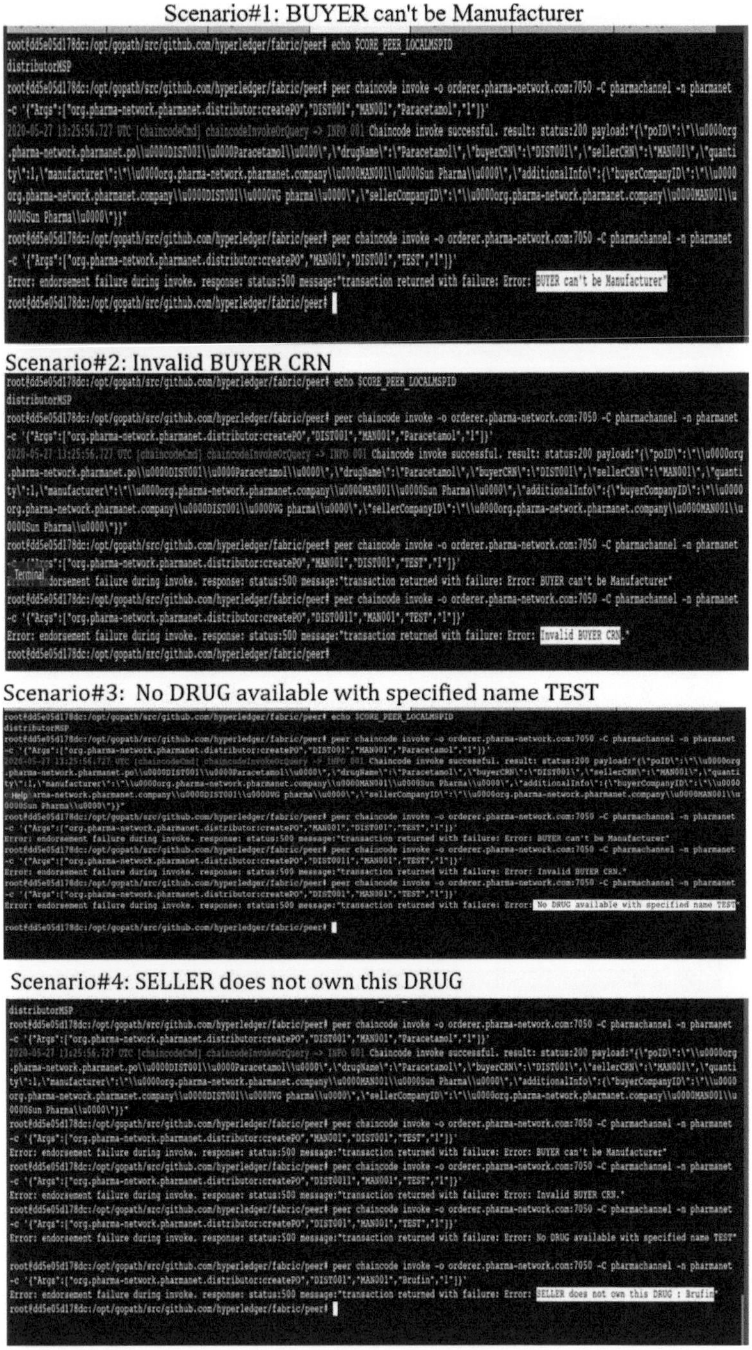

Scenario#1: BUYER can't be Manufacturer

Scenario#2: Invalid BUYER CRN

Scenario#3: No DRUG available with specified name TEST

Scenario#4: SELLER does not own this DRUG

Fig. 11 A createPO() query invoking chaincode through CLI

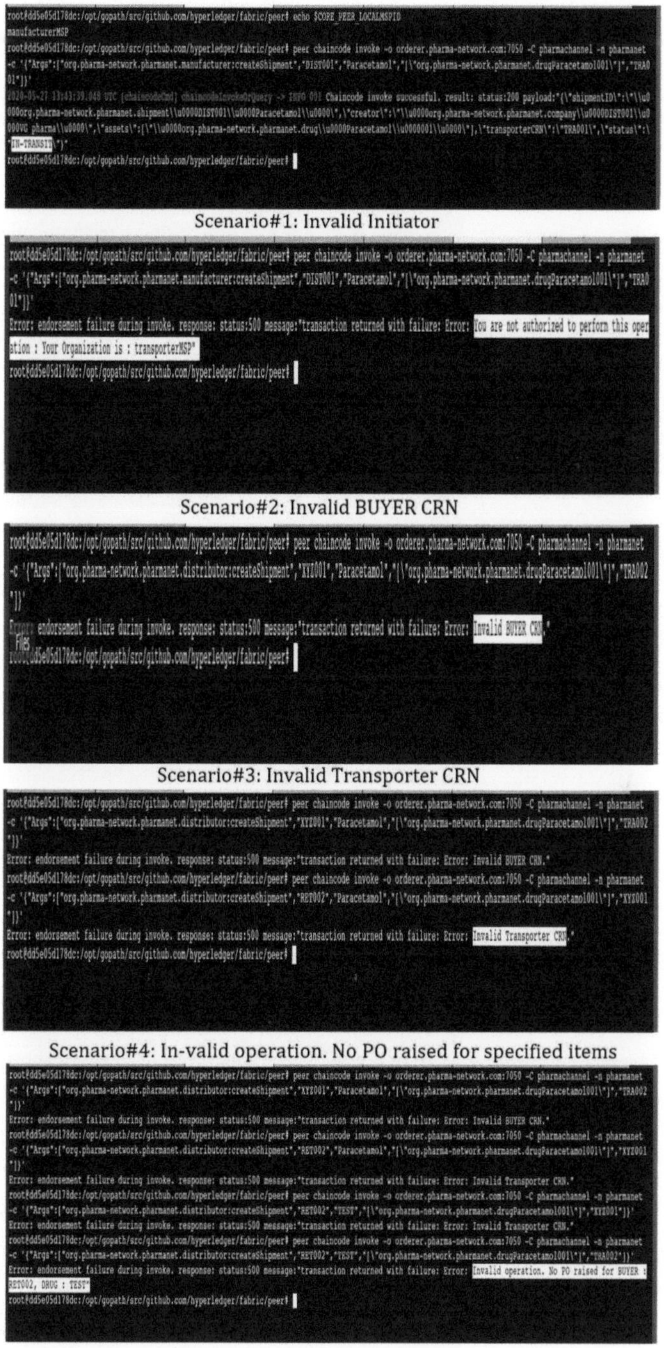

Scenario#1: Invalid Initiator

Scenario#2: Invalid BUYER CRN

Scenario#3: Invalid Transporter CRN

Scenario#4: In-valid operation. No PO raised for specified items

Fig. 12 A createShipment() query invoking chaincode through CLI

Fig. 13 A updateShipment() query invoking chaincode through CLI and Postman

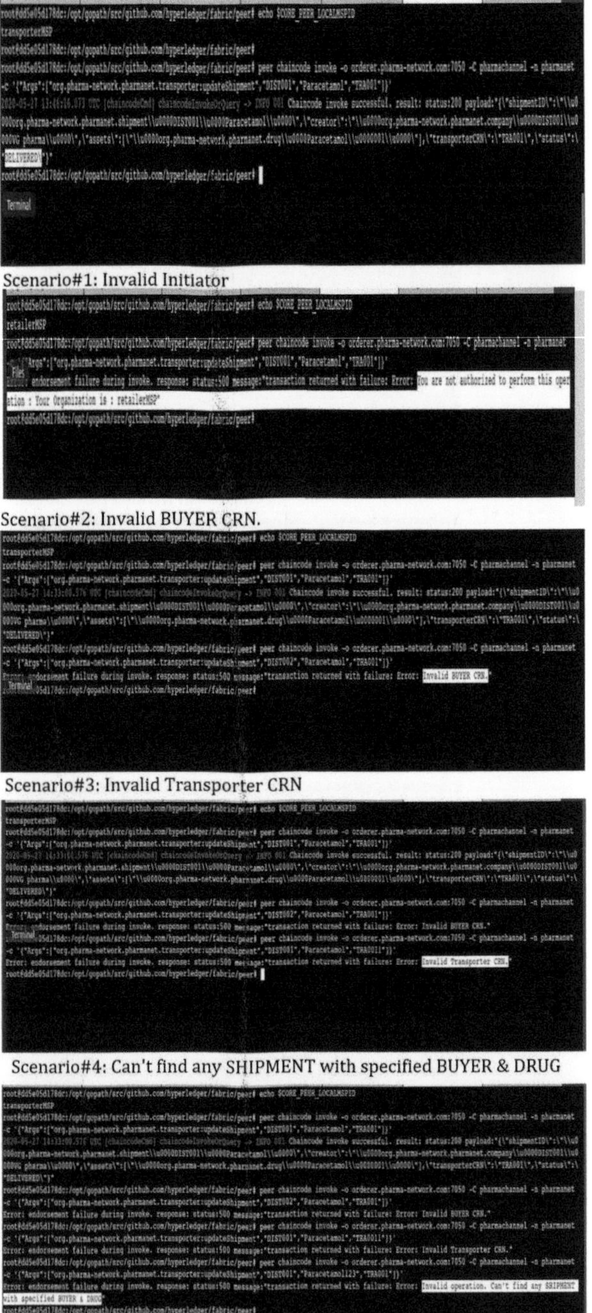

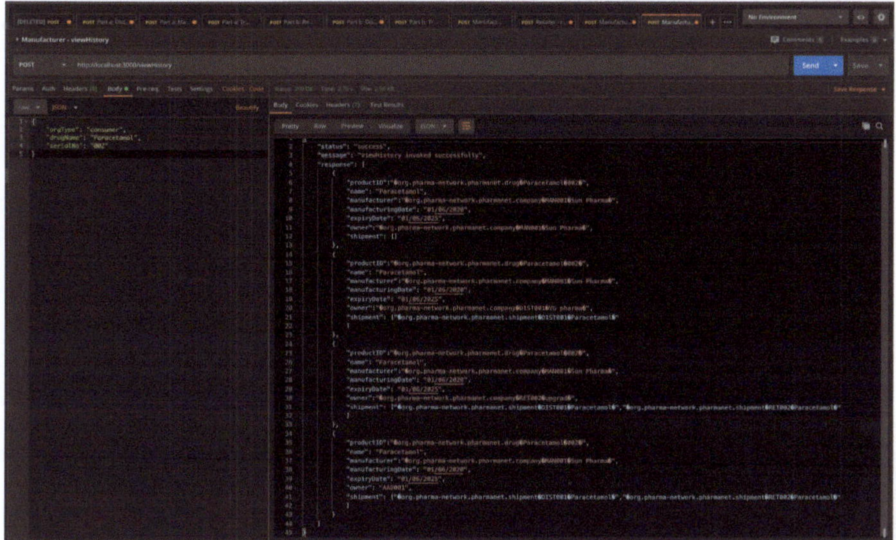

Fig. 14 A viewHistory() invoked by the manufacturer

Table 6 The required software for setting up Hyperledger Explorer environment

Required software	Version description
Nodejs	8.11.x
PostgreSQL (open source relational database system)	9.5 greater
Jq (command-line JSON processor)	
Docker	19.03.1
Docker Compose	1.23.2

6 Real-Time Visualization of Blockchain Operations Using Hyperledger Explorer

The Hyperledger Explorer [44, 45] is a web-based interface to visualize the operations of blockchain network. This tool supports many blockchain solutions like Hyperledger Fabric and Hyperledger Iroha. We have configured the Hyperledger Explorer with PSC fabric to monitor the operations of the network. The software listed in Table 6 are required to set up the environment for Hyperledger Explorer.

Blockchain Explorer helps us to visualize and monitor the following network parameters:

- Latest block details, transactions, and chaincode
- View blocks and transactions
- Blocks and transactions metrics by hour and minute

- Search and filter blocks, transaction by date range, and channels

6.1 Blockchain Explorer Output Screens

- *Dashboard.* Figure 15 shows the Blockchain Explorer dashboard which shows the latest activity in the Hyperledger PSC network. In the landing page of the Blockchain Explorer, we can see the navigation tabs, namely, DASHBOARD, NETWORK, BLOCKS, TRANSACTION, CHAINCODE, and CHANNELS. In DASHBORAD window, we can see information such as number of blocks created, number of transactions in the ledger, number of nodes in the network, and chaincode of the network. Also we can see the block detail which is in this case Block 5. This includes the following details: (1) channel name: pharma channel; (2) data hash; and (3) number of transactions. Also visible on the metric panel are metrics like BLOCKS/HOUR, BLOCKS/MIN, TX/HOUR, and TX/MIN.

In Fig. 16, it shows the view of latest dashboard after adding one more block into PSC blockchain fabric network in recent times. And it shows that the latest block contains four transactions in it. Also, we can read the information about the block, i.e., the block creation time from metric panel.

- *Network.* Blockchain Explorer stores the pharma channel's properties as tables under the network tab. Property names like Peer Name, Requested URL, Peer Type, MSPID, High, Low, and Un-signed are displayed in the column values of the table. The row values in Fig. 17 include the kind, Membership Id -MSPID, Ledger Height, and Un-signed. They also include the number of peers added into each organization of the PSC fabric network.

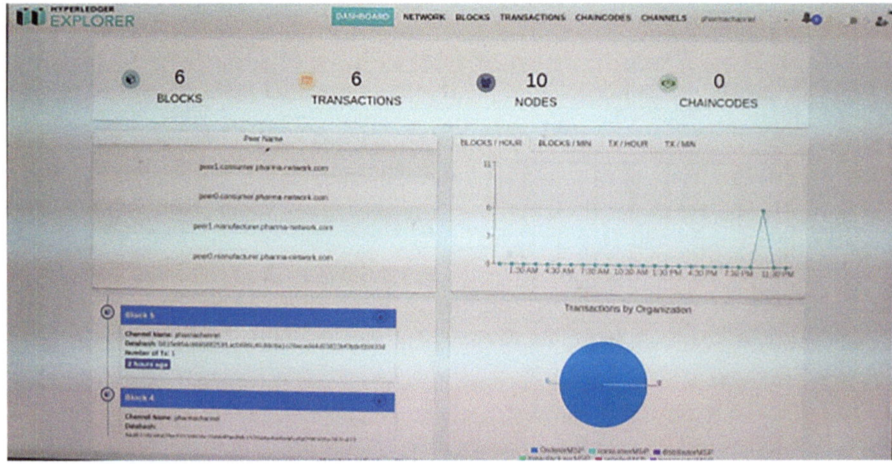

Fig. 15 Blockchain Explorer dashboard

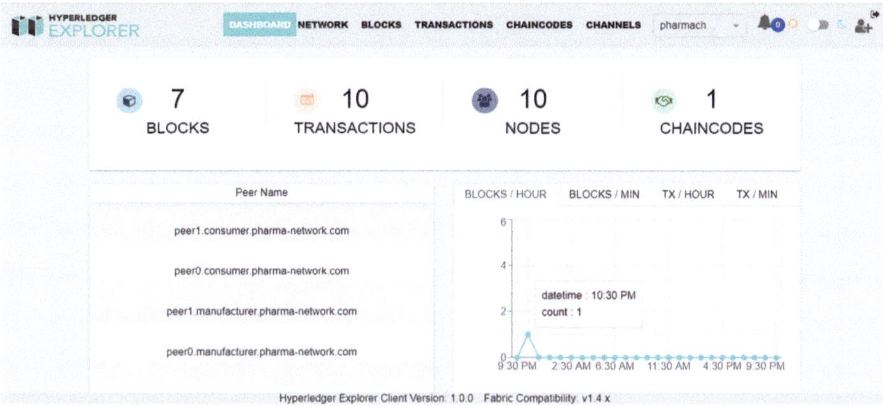

Fig. 16 Updated dashboard of Blockchain Explorer

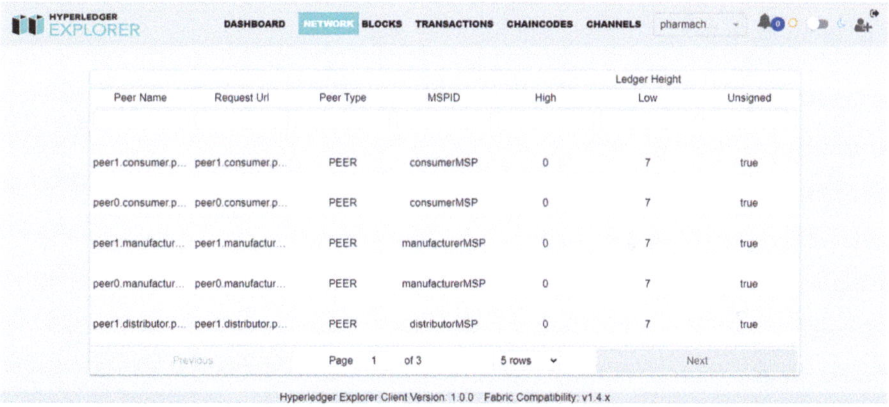

Fig. 17 Network tab

- *Blocks.* The blocks tab in Fig. 18 stores the details of the individual block such as block number, channel name, number of transactions, block hash, data hash, and previous hash. By clicking the block hash from the window, we get to see the individual block details which is mentioned in Fig. 19.

- *Transactions.* This window shows the information about transactions performed by the node in PSC fabric network. The column values of the table are block Creator, Channel Name, Tx Id, Type, Chaincode, and Timestamp. The row values shown in Fig. 20 indicate that the manufacturerMSP initiates the transaction by instantiating the chaincode. We can also read the time stamp details of the corresponding transaction.

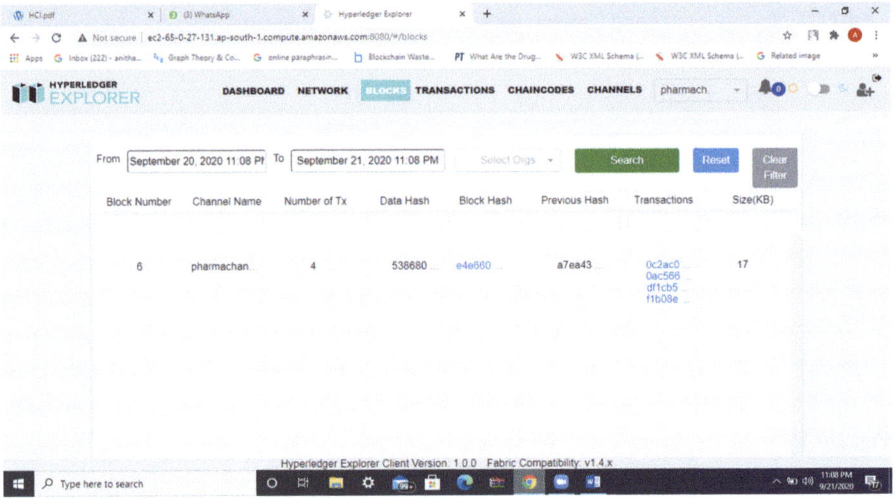

Fig. 18 Blocks tab

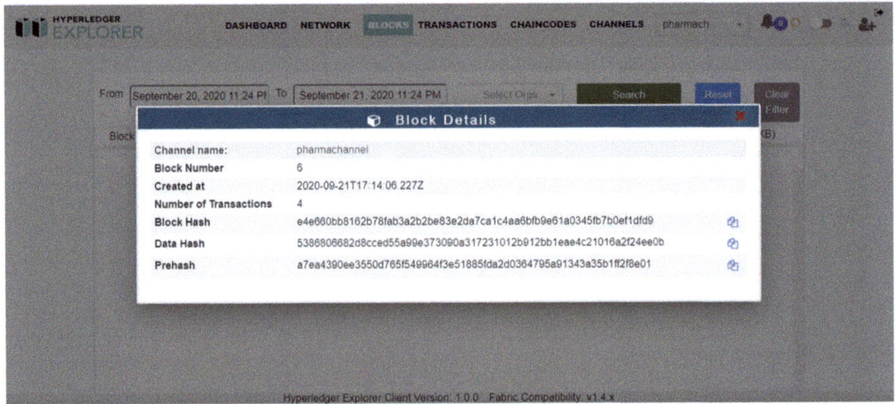

Fig. 19 Block details

- *Chaincode.* The tab shown in Fig. 21 displays the details of chaincode such as Chaincode Name, Channel Name, Path, Transaction Count, and Version. Here the row values are pharmanet, pharma channel, /opt/gopath/src/github.com/ hyperledger/peer, 7, and 1.2, respectively.

- *Channels.* The tab in Fig. 22 shows the details of PSC fabric network channel such as ID, Channel Name, Blocks, Transactions, and Timestamp.

Table 7 compares our proposed work with other existing solutions mentioned in the literature section.

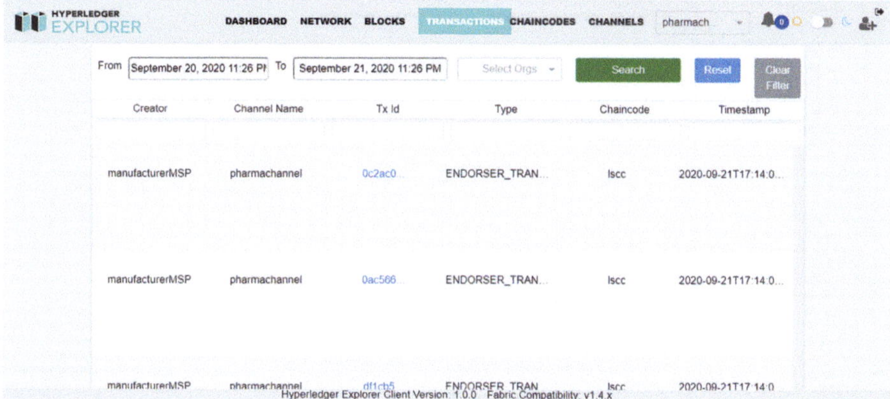

Fig. 20 Transaction tab

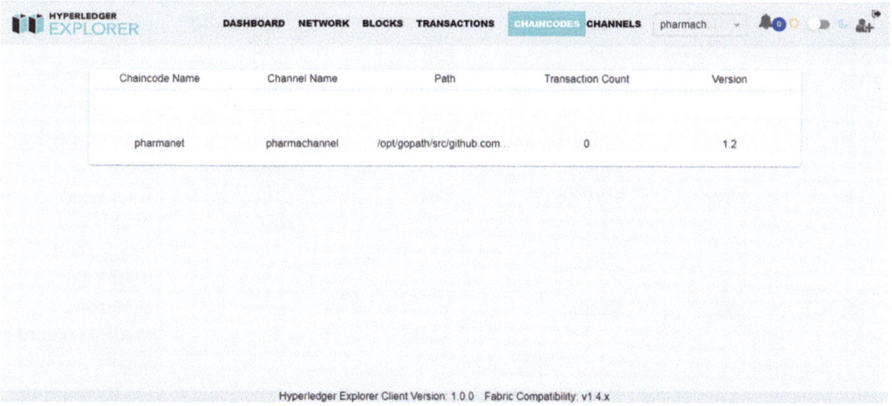

Fig. 21 Chaincode tab

Our suggested solution uses the Hyperledger fabric technology to create a private blockchain network with authorized nodes in order to communicate medicine information and also makes it easier to spot fake medications while a drug is being delivered. Using Hyperledger Explorer, the network parameters for our proposed model, such as the number of blocks and tractions, nodes, and chaincode, have also been tracked and shown. This clarifies our suggested system's visibility and operational mode.

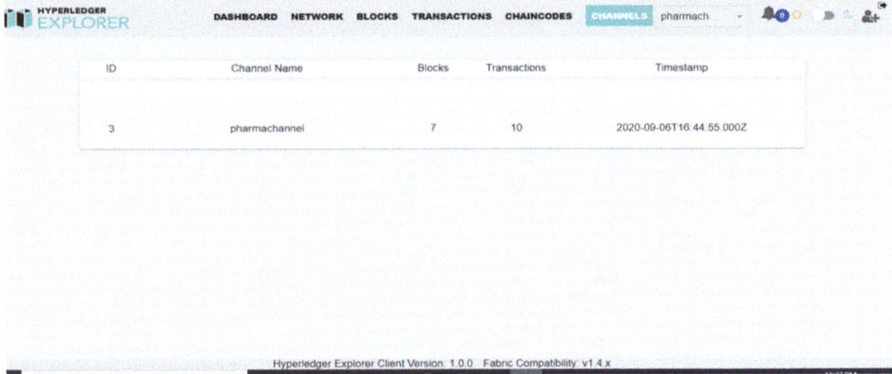

Fig. 22 Channels tab

Table 7 Comparison of a proposed work with an existing solution

Existing solution	Smart contract	Implementation type	Consensus algorithm	Technology used	Purpose
[32]	Yes	Public	Yes	Ethereum blockchain	Drug SCM
[33]	Yes	Permissioned public	Yes	Ethereum blockchain	Patient medical data
[34]	Yes	Private	Yes	Ethereum ERC20	Integrated healthcare blockchain platform
[35, 36]	Yes	Public	Yes—proof-of-work		Electronic medical record EMR
Blockchain-based proposed PSC system	Chaincode	Permissioned private	Yes—practical Byzantine fault tolerance (PBFT)	Hyperledger fabric	(a) Identifying the counterfeit drugs and eliminating them (b) Monitoring and visualizing the network parameters

7   Conclusion

A permissioned private blockchain PSC network with smart contract chaincode and a distributed digital ledger has been built using Hyperledger fabric. The method increases confidence by preventing the introduction of fake drug data into the system and giving participants complete transparency. Also, we have set up Blockchain Explorer to track and display the PSC network. One can view operational metrics with this tool, such as the number of blocks made per hour, blocks created per minute, transactions created per hour, transactions generated per minute, etc. This comprehensive solution might be further developed to expand the network's size and performance and deploy the system in a live setting for actual people to use.

References

1. A white paper on "Pharmaceutical serialization: Compliance and beyond". Available Online: https://www.ey.com/Publication/vwLUAssets/ey-pharmaceutical-serialization-compliance-and-beyond/%24FILE/ey-pharmaceutical-serialization-compliance-and-beyond.pdf
2. Shah, N. A., et al. (2015). Counterfeit drugs in India: Significance and impact on pharmacovigilance. *International Journal of Research in Medical Sciences, 3.* https://doi.org/10.18203/2320-6012.ijrms20150596
3. A Counterfeit problem in Africa. Available Online: https://www.youtube.com/watch?v=11Z4-XYoZAE
4. India, China are leading sources of counterfeit medicines: Report. Available Online: https://www.theweek.in/news/biz-tech/2019/05/04/India-China-are-leading-sources-of-counterfeit-medicines-report.html
5. A white paper on "India's pharma supply chain: Does the industry have what it takes to win", A.T. Kearney (2016). Available Online: https://www.indiaoppi.com/wp-content/uploads/2019/12/Indias-Pharma-Supply-Chain-Does-the-Industry-Have-What-It-Takes-to-Win.pdf
6. A white paper on "Deloitte's 2019 global blockchain survey". Available Online: www.deloitte.com.
7. Swan, M. (2015). *Blockchain: Blueprint for a new economy*. O'Reilly Media, Inc..
8. Ismail, L., & Materwala, H. (2019). A review of blockchain architecture and consensus protocols: Use cases, challenges, and solutions. *Symmetry, 11*, 1198.
9. Joshi, A. P., et al. (2018). *A survey on security and privacy issues of Blockchain technology*. Mathematical Foundation of Computing, American Institute of Mathematical Sciences.
10. Gupta, A., et al. (2014). Cryptography algorithm: A review. *International Journal of Engineering Development and Research, 2*(2), 1667.
11. Massessi, D. Public Vs private blockchain in a nutshell. Available online: https://medium.com/coinmonks/public-vs-private-blockchain-in-a-nutshell-c9fe284fa39f
12. Zhang, S., & Lee, J.-H. (2019). *Analysis of main consensus protocols of blockchain*. Science Direct.
13. Nakamoto, S. (2008). Bitcoin: A peer-to-peer electronic cash system. Available: https://bitcoin.org/bitcoin.pdf
14. Tschorsch, F., & Scheuermann, B. (2016). Bitcoin and beyond: A technical survey on decentralized digital currencies. *IEEE Communications Surveys & Tutorials, 18*(3), 2084.
15. Wood, G. Ethereum: A secure decentralized generalized transaction ledger. Available Online: https://gavwood.com/paper.pdf
16. A white paper on "Hyperledger fabric framework". Available online: https://hyperledger-fabric.readthedocs.io/
17. A white paper on "An introduction to hyperledger". Available online: https://www.ibm.com/downloads/cas/0XMOQJNP

18. A white paper on "Hyperledger architecture volume 1". Available on line: https://www.hyperledger.org/wp-content/uploads/2017/08/Hyperledger_Arch_WG_Paper_1_Consensus.pdf
19. Sukhwani, H. (2018). *Performance modeling & analysis of hyperledger fabric (performance blockchain network)*. Duke University.
20. A white paper on "Corda: A distributed ledger". Available Online: https://www.r3.com/wp-content/uploads/2019/08/corda-technical-whitepaper-August-29-2019.pdf
21. Kapoor, D., Vyas, R. B., & Dadarwal, D. (2018). An overview on pharmaceutical supply chain: A next step towards good manufacturing practice. *Drug Designing & Intellectual Properties International Journal, 1*, 107.
22. Shah, N. A., et al. (2015). Counterfeit drugs in India: Significance and impact on pharmacovigilance. *International Journal of Research in Medicine Sciences.*
23. O'Hagan, A., & Garlington, A. (2018). Counterfeit drugs and the online pharmaceutical trade, a threat to public safety. *Forensic Research & Criminology International Journal, 6*(3), 151–158.
24. Saurabh, V., et al. (2014). The business of counterfeit drugs in India: A critical evaluation. *International Journal of Management and International Business Studies, 4*(2), 141.
25. Khan, A. N., & Khar, R. K. (2015). Current scenario of spurious and substandard medicines in India: A systematic review. *Indian Journal of Pharmaceutical Sciences, 77*, 2. Available Online: https://www.ncbi.nlm.nih.gov/pmc/articles/PMC4355878/?report=reader#!po=38.2353
26. Gupta, P., et al. (2012). Counterfeit (fake) drugs & new technologies to identify it in India. *International Journal of Pharmaceutical Sciences and Research, 3*, 4057.
27. Implementing India's Drug serialization and traceability requirements to advance patient safety and support global trade. DAVA.
28. Siyal, A. A., et al. (January 2019). *Application of blockchain technology in medicine and healthcare: Challenges and future perspective*. MDPI.
29. Faisal Jamil et al, "A novel medical blockchain model for drug supply chain integrity management in a smart hospital", MDPI, 2019.
30. A white paper on "Over coming barriers to next gen supply chain innovation". The 2018 MHI annual industry report. https://www.supplychain247.com/paper/2018_mhi_annual_industry_report_overcoming_barriers_to_nextgen_supply_chain
31. Raj, N. H. Writing your first simple Hyperledger Fabric Chaincode in Go. Available Online: https://www.skcript.com/svr/writing-your-first-simple-hyperledger-fabric-chaincode-in-go/
32. The MediLedger Project 2107 Progress Report, Whitepaper. Available online: https://uploads-ssl.webflow.com/59f37d05831e85000160b9b4/5aaadbf85eb6cd21e9f0a73b_MediLedger%202017%20Progress%20Report.pdf
33. Medicalchain Whitepaper, 2018. Available online: https://medicalchain.com/Medicalchain-Whitepaper-EN.pdf
34. MeFy, MeFy Whitepaper, 2018. Available online: https://icosbull.com/whitepapers/3576/MeFy_whitepaper.pdf
35. Azaria, A., et al. (2016). MedRec: Using blockchain for medical data access and permission management. In *International conference on open and big data.*
36. Ekblaw, A., Azaria, A., Halamka, J. D., & Lippman, A. (2016). A case study for blockchain in healthcare: "MedRec" prototype for electronic health records and medical research data. In *Proceedings of the IEEE Open & Big Data Conference, Vienna, Austria, 22–24 August 2016* (Vol. 13, p. 13).

Quantum and Blockchain for Sustainable Healthcare Ecosystem

Syed Muzammil Munawar, Dhandayuthabani Rajendiran, and Khaleel Basha Sabjan

Abstract Electronic health records (EHRs) have been replacing paper-based medical records in the majority of healthcare facilities. The present EHR frameworks, however, have difficulties with management, trustworthiness, and secure data storage. In the healthcare industry, user personal data ownership and interoperability are critical challenges. Although blockchain technology provides various benefits like security, stored records protection, and immutability, its potential usage in EHR systems is still not well appreciated. This project intends to bridge this knowledge gap by developing an interoperable EHR architecture based on blockchain technology that complies with numerous national and international EHR standards like the Health Insurance Portability and Accountability Act (HIPAA) and Health Level 7 (HL7). This chapter describes the problems associated with interoperability of EHR based blockchain frameworks, evaluates the diverse national and international EHR standards, and outlines the EHR standard requirements for the interoperability. The suggested architecture has the potential to offer immutability, user-controlled record management without the need for centralized storage and security, and user control over stored records without the requirement for centralized storage, as well as safer ways to exchange health information for the healthcare industry. The work's contributions encompass advancing knowledge of the possible application of EHR based blockchain framework and presenting interoperable framework capable of satisfying the EHR standards. In summary, this chapter can enhance the secure transmission and storage of EHR while safeguarding the confidentiality, accuracy of medical information, and privacy, thereby holding significant ramifications for the healthcare sector.

Keywords Blockchain technology · Electronic health records · Digital innovation

S. M. Munawar (✉) · D. Rajendiran · K. B. Sabjan
Department of Biochemistry, C. Abdul Hakeem College (Autonomous),
Ranipet, Tamil Nadu, India

© The Author(s), under exclusive license to Springer Nature
Switzerland AG 2024
S. Pulipeti et al. (eds.), *Quantum and Blockchain-based Next Generation*
Sustainable Computing, Contributions to Environmental Sciences & Innovative
Business Technology, https://doi.org/10.1007/978-3-031-58068-0_5

1 Introduction

Blockchain-based projects have gained popularity in recent years. However, blockchain is a relatively new and complex technology, and it would be incorrect to believe that it can be applied immediately or that adjustments can be made with ease [1]. Blockchain is not a panacea for today's medical record management issues. However, it presents some potential to enhance the current system, which is why it is both fascinating and challenging to study. After being used for several years as the foundation of a digital currency, the disruptive technology known as blockchain is now emerging as an open source solution with potential applications across various fields. Due to the built-in cryptography and decentralization, there are creative uses of blockchain technology in healthcare. This technology's appeal lies in its capacity to scale a distributed scheme that ensures data integrity and preserves confidentiality. It strengthens the security of EHRs, facilitates the monetization of health data, enhances interoperability among health organizations, and contributes to the battle against counterfeit medications. Blockchain technology has the potential to revolutionize numerous sectors within healthcare. Especially, digital contracts enabled smart contracts removing intermediaries from the payment chain; smart contracts will minimize costs [2, 3]. The viability of blockchain in healthcare predominantly hinges on the integration of related advanced technologies within the ecosystem. These include systems monitoring, health insurance, pharmacovigilance, and clinical trials. With the blockchain platform, hospitals can even map their services over the entire life cycle using device tracking. The effectiveness of blockchain technology can enhance medical history management, especially insurance tracking and brokerage process, accelerating clinical activities through streamlined data maintenance. Overall, the technology would greatly improve and eventually revolutionize the way patients and physicians process and use medical records and enhances the healthcare services quality. Blockchain offers both the private and public sectors previously unheard of opportunities. Any organization that harnesses these technologies stands to significantly optimize and modernize existing processes, develop entirely new business models, and deliver innovative products and services to a new generation of consumers. This is not a utopian, tech-driven vision; it is now feasible to maintain an indisputable record of every transaction and eliminate the intermediaries for digital transactions [4, 5]. All industrial, social, and economic sectors will experience accelerated process, transaction transparency at real time, and reduced costs as a result. Therefore, current strategy focuses on the blockchain technology in the healthcare industry, by evaluating its key attributes and categorizing the various blockchain types that are currently in use; the blockchain is investigated. The healthcare industry was picked specifically because, despite being still extremely varied, it is one of the primary sectors of digitization; because of this, the chapter examines instances in which information systems enabling the management and preservation of patient records in the National Health System (NHS) appear to be completely incapable or nonexistent. Patients frequently overlook tests they've undergone, struggle to access specific clinical data when required, or opt against seeking second opinion due to practical impracticalities. The chapter proposes a remedy to these issues in order to strengthen the NHS, which is ineffective that it seems to erode public confidence, leading more Italians to resort to private healthcare. Additional objective is to establish a decentralized database model for storage of personal medical data, accessible conveniently through a smartphone at anytime, from anywhere. However, it's important to recognize the intricacy of this novel approach, which, though still in its early development stages, has clearly divided opinion on whether to use it or not. Introducing a highly novel and complicated IT system into the overhaul of a system like *Sistema Sanitario Nazionale* (SNN) could have two significant implications. On one hand, integrating this technology would bring numerous benefits. Nevertheless, some users might resist substantial changes as they can be challenging for certain individuals to embrace. An approved proof of authority (PoA) blockchain is the solution suggested in this chapter for managing and archiving registered patients' electronic medical records. The fact that the data being handled is personal led to the permissioned type of blockchain being selected. This method ensures efficiency for both doctors and patients and, ideally, restores faith in the *Sistema Sanitario Nazionale* (SSN). It ensures immutability and transparency that are essential for storage and secure management [6, 7]. Such a model would need cooperation and standardization among stakeholders, who most likely have different demands and interests, in order to reach its full potential. Furthermore, one might question whether blockchain is merely another instance of suboptimal healthcare innovation, considering the application of blockchain technology to tackle these issues in the SSN and the substantial costs it would entail. The optimum course of action is seen to be the gradual introduction of this technology. Tests and analyses will determine the optimal course of action to ensure gradual and sustainable increase in returns on the technology investment. It should be emphasized that the goal of this chapter is to build the foundation for future study that will examine cost-benefit analysis and metrics like return on investment to assess the viability of the project. This chapter restricts itself to outlining a fundamental model, the ramifications of which should be further investigated. If such a large-scale endeavor seems intimidating, consider the world before the blockchain, when no one would have considered implementing interoperable trust architecture in the healthcare industry. Blockchain technology holds an advantage for the advancement of this system due to its divergence from previous approaches, as it has the ability to streamline the transfer of medical data through an immutable and decentralized architecture, offering a secure, unalterable, and collectively accessible registry of medical records. In reality, the development of new citizen services can be accelerated by blockchain technology. Politicians need to take the initiative, fully comprehend the basic significance of innovation in public and healthcare systems, and encourage digital culture and related infrastructure [8, 9]. Transparency, timeliness, accuracy, and security are implicit criteria that rise when numerous parties are involved in a transaction, as is the situation with public health services. Governments must fulfill these needs while maintaining a high standard of quality if they want to add value to their citizens' lives and earn their trust. A system like blockchain, engineered to ensure the precision and openness of transactions and the secure handling and preservation of data, unquestionably meets all the necessary criteria to earn the privilege of reimagining administrative processes and land management, especially in a period when trust in public institutions is severely diminished. The development and growth strategy's implementation comprises numerous decision-making levels including, but not limited to, businesses, academic institutions, people, start-ups, research facilities, etc. and emphasizes a more participative management approach for regional systems. Hence, a robust digital infrastructure is imperative to harness the possibilities brought forth by emerging technologies such as blockchain, artificial intelligence, or the "Internet of Things." This infrastructure forms the cornerstone of fostering a resilient "IT" culture rooted in consistent and reliable engagement with the general public, businesses, and national authorities. A tangible document is one of the many difficulties that arise as a result of the blockchain technology implementation in healthcare, which is undoubtedly a challenging and multistep procedure. Psychological barriers to coping with change, even when it is essential or beneficial, are another factor contributing to the psychological holdup of technical change. When dealing with an innovation that is out of the ordinary and familiar, it's common to feel uneasy and insecure. But resistance to change shouldn't stop the inescapable digitalization and innovation process. In conclusion, blockchain technology has the potential to significantly expand the horizons and development scenarios for public services. The decentralization of public administration will become an inherent, inevitable, and groundbreaking outcome of leveraging this technology once it reaches a sufficient level of technological maturity. Therefore, it's crucial to encourage governments to take this course and implement the basic adjustments that all citizens demand. In conclusion, blockchain technology has a lot of promise and might bring about a lot of advantages. Blockchain research is still in its early stages, but the application of technology can improve societal integrity and boost public institution trust among the populace. It has various restrictions because it is still in an early stage. We are currently in the initial and exploratory analysis phase of our study on the use of healthcare based blockchain, which takes a qualitative rather than a quantitative approach. Our goal is to continue our research to support the use of a standardized and exposed framework for the monitoring and verification of EHRs. Blockchain is unquestionably revolutionary, and as a result, many of its consequences have only been theorized about. We seek to establish a pricing model appropriate for the health realm as well as assess how NHSs can adapt to such digital divide advancements in our future study. The scalability and efficiency of the process over time that affect not only process itself but also other processes that might save additional resources and time are our goal rather than merely cost reductions. Additionally, the treatment of human health is significantly influenced by the climate [10].

1.1 Blockchain

Blockchain is increasingly being used in the healthcare sector owing to its essential encryption and decentralization that complement the protection of patients' digital scientific information, with the blockchain age, for example, different health areas can alternate in regions. In healthcare, virtual contracts definitely represent one of the block applications from the highest critical chain [11]. Away with the middlemen of the price chain, smart contracts will reduce costs of blockchain capacity in the health sector; it is particularly important to adopt the associated cutting-edge technology ecosystem, including device tracking, healthcare, medical surveillance, and research. Hospitals can follow their services using the blockchain framework even throughout the life cycle by tool tracking [12]. The blockchain age can be used properly to empower the data subject documentation management, in particular monitoring of coverage and the mediation process; this accelerates medical operations with optimized information storage. The blockchain is a set of blocks that can be encrypted using community nodes and fixed ledger operations. Events/transactions are block by block created at its timestamp which offers sign in with unique creation date and time to ensure authenticity. An integral part of the generation of the blockchain, everything is related to encryption and irreversibility or hashing rule set, in a one-way algorithmic function. The blockchain displays completed transactions in a block that is linked to a previous block by its hash. The hash produces a set of rules that take into account the preceding block in the chain. A block becomes an integral part of the unchangeable blockchain, also known as the ledger or an encrypted file, as soon as possible as it is finished and is then safely kept [13]. For example, a new block must be confirmed, validated, and encrypted before it can be added to the blockchain. You need to make a brand novel precise botch by a set of hash system that conceals all block statistics to defend from people who surely execute, as this process is known as miner. As seen above, it is difficult to find a sole definition of blockchain for the types of approaches and sectors where it can be implemented. Of course, this can be viewed as a transactional database—a secure way to reassign stats to encrypted blocks. Blockchain can also be seen as an evolution of the idea of the ledger, the filing cabinet that is constantly updated. Finally, the blockchain can be interpreted as public documents that are available for everyone to inspect, and transparency is part of its base houses of this generation [14].

2 Reviewing Blockchain Technology

2.1 Blockchain Technology Implementation

The so-called owing ledger gave rise to the blockchain. The design of the devoted ledger is the result of the extension of the national ledger, which then proceeds to expand the decentralized ledger. At first, national ordinary intelligence represents the ledger as a one-to-many association in which the whole group of people is governed by the possessions of a solitary power to which every lump referred. This defined version of the ledger is accepted as true of in an isolated checkpoint in the middle of the community. First a crucial step was taken with the transition from a centralized to decentralized version of the ledger; even as in the centralized version, it was not the easiest an authority that has prepared many municipalities in the decentralized version corresponding governments are formed, each reflecting a one-to-many relationship. The genuine evolution happens with the chosen ledger description, which embraces selected high-quality decision, and the business venture of the society is constructed around the product novel design of mutual belief that among all the participants, no one has better power than anybody as well [15].

2.2 Quantum Computing

These qubits inherit two crucial quantum traits, superposition and entanglement, because they may be represented by any quantum particle. Like a regular bit, a qubit can be in the states of 0 and 1, but it can also be in intermediate states where each state has a probability. Superposition is the name of this condition, and while a qubit is in superposition, its state is unknown, allowing a quantum computer to operate simultaneously in both states. This lasts until the measurement of the qubit in a controlled state causes its quantum state to collapse to 0 or 1. Qubit pairs that are entangled, or in which there is a correlation between the qubits, can also be produced by scientists because of the state of one member to remotely affect the state of the other member. When two qubits are entangled, they are in the same quantum state and are therefore considered to be one object with four separate states [15]. When the quantity of qubits doubles, these two phenomena generate an exponential rise in the performance of a quantum computer. The tasks that our computers appear to be able to complete in parallel are actually just divided into their basic operations, which are then carried out in one operation at a time while switching between jobs [15]. A quantum computer may calculate all possible states described by 2n, where n is the number of qubits, simultaneously by processing numerous inputs in parallel using entanglement and superposition [16]. The qubits' quantum state is extremely unstable, which makes it difficult to shield quantum computers from ambient noise or disturbances of any type, including vibrations, temperature changes, and even interactions between the qubits themselves. Quantum computers must therefore be stored in extremely secluded locations with temperatures near to zero degrees. These effects are referred to as decoherence when they are discussed in relation to quantum computers. These qubits inherit two crucial quantum traits, superposition and entanglement, because they may be represented by any quantum particle. Like a regular bit, a qubit can be in the states of 0 and 1, but it can also be in intermediate states where each state has a probability. Superposition is the name of this condition, and while a qubit is in superposition, its state is unknown, allowing a quantum computer to operate simultaneously in both states. This lasts until the measurement of the qubit in a controlled state causes its quantum state to collapse to 0 or 1. Qubit pairs that are entangled, or in which there is a correlation between the qubits, can also be produced by scientists because of the state of one member to remotely affect the state of the other member. When two qubits are entangled, they are in the same quantum state and are therefore considered to be one object with four separate states. When the quantity of qubits doubles, these two phenomena generate an exponential rise in the performance of a quantum computer. Despite the fact that our computers appear to be able to accomplish activities in parallel, they are actually just divided into basic operations and then carried out one operation at a time while switching between jobs. A quantum computer may calculate all possible states described by 2n, where n is the number of qubits, simultaneously by processing numerous inputs in parallel using entanglement and superposition [17]. The qubits' quantum state is extremely unstable, which makes it difficult to shield quantum computers from ambient noise or disturbances of any type, including vibrations, temperature changes, and even interactions between the qubits themselves. Quantum computers must therefore be stored in extremely secluded locations with temperatures near to zero degrees. These effects are referred to as decoherence when they are discussed in relation to quantum computers. This causes information loss or a change in the quantum state of qubits. The answer is to encode a logical qubit into several physical qubits; the system may then detect faults and partially repair them as a result of the ensuing redundancy. In exchange for a significantly higher requirement for qubits, this extends the time before errors start to accumulate during calculations. However, quantum computers do not have the same level of parallelism as graphics cards, which may create cores and perform simultaneous execution tasks. Therefore, it is likely that quantum computers will be utilized to solve hard and well-defined problems rather than to replace CPUs and GPUs.

3 Blockchain Types

The identification and classification of the many categories of blockchains is of great importance as this is where they reach their maximum. You can choose the right blockchain type for your world and oversight survey. Many blockchains can be broken down into major types, unauthorized and trusted blockchains. The biggest change is that Bitcoin is no longer functional; you can no longer make transactions or add new blocks without first receiving approval from the community. The form extends beyond the road of truth that there is no power to govern access when there are nonregulatory conditions that permit perfect decentralization of these blockchains. A form of blockchain known as a public blockchain notifies all community nodes of any modifications to the chain roughly at the same time, confidentiality. Last but not least, unofficial blockchains are also employed, such as large-scale statistical databases that need to be consulted and maintain consistency over time. Contrarily, permitted blockchains function first, are totally based on a very specific instance, and offer a crucial foundation for making decisions that may also be seen by the community and used by a member of the community. However, even if all nodes had access to the blockchain and could see modifications, the substance of each block's statistics was encrypted to provide a high enough level of security. Instead of letting everyone participate without exception, when verifying transaction process, some trusted nodes can be fair by using critical framework to do this work; therefore, trusted blockchains are used, idea of community governance and centralization [18]. Private blockchains provide a wide range of proportional functions. For blockchains that have been white-listed, these are hidden private networks that are governed by authorities who have the authority to decide who else may and cannot be white-listed in community data [19]. The instructions for use can also be changed if the blockchain organization so desires, and 51% attack/harvesting is unlikely. Miners who control more than 50% of a community's hash price or processing power may be disqualified. Minors are recognized. Transaction costs are lower. Better are hyperlinks that connect nodes. The aforementioned blockchain types can also be merged as shown below:

3.1 Unrestricted Public Blockchain

There isn't a place to enter transactions or verify documents. Minors are anonymous and are therefore considered unreliable. Access may be provided in certain circumstances on a publicly authorized blockchain, at the sole discretion of the blockchain business. One near variation of humans is the miner. The most authorized and authenticated individuals can join the community as this type of blockchain operates on a large and limited perimeter recognized by all. Therefore, minors are considered a trusted account [20].

4 Blockchain in Healthcare and Their Needs

Just like in healthcare interest groups, the urgent need for improvement will move at an exceptionally high speed. Today, fun, empowered fitness clubs are in demand of best and newest technologies. Blockchain plays a vital role in the rehabilitation of the healthcare realm. Moreover, the landscape of the so-called fitness gadget moves toward a person-targeted method specializing in the most important thing:

realizable services and adequate sources of health care in all respects. The blockchain helps the health authorities to provide the data subject with sufficient and excessive care. The exchange of health data has ended again intense, and tedious system that hints to overwork quickly healed costs for the use of this generation. Using the blockchain generations, residents can also take part in fitness classes and look at the programs. Also, higher education and general facts about the common good will embellish the medicine of famous communities. There is a central database used to manipulate the entire health device and Corporation [21]; until now, the most relevant concerns are protection of facts, sharing, and interoperability within population adaptation management; with full blockchain security, fact sharing, interoperability, integrity, and real-time updates and access, this particular issue is foolproof when done right. Serious problems arise in particular with regard to the protection of facts particularly in the field of personalized medicines and wearable devices. The sick and academics need an easy and secure way to store, transfer, and fact check on networks with no security issues; then blockchain generation is performed to solve these problems [22].

5 Revolutionizing Healthcare: Exploring the Blockchain Technology Capabilities

In healthcare, blockchain has a wide variety of packages and features [23]. The creation of the ledger helps the health department scientists find the genetic code and make permanent changes possible in the clinical facts of the interested party, the drug supply chain management facilitating the secure modification of clinical data about the data subject. Figures 1, 2, 3, and 4. A comprehensive range of essential features and enablers of blockchain technology in various areas of healthcare and related fields are depicted, "shows the range of necessary features and enablers of the blockchain philosophy in myriad areas of health and related fields." The blockchain technology offers a wide range of features and enablers that can revolutionize the healthcare industry. From health protection statistics to clinical facts, interoperability, surveillance, and digital explosions, the blockchain generation provides amazing derivatives and functions that can extend and practice healthcare services

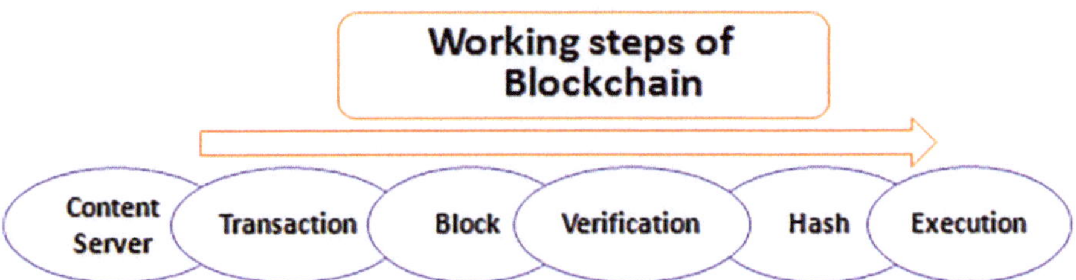

Fig. 1 Steps involved in the functioning of blockchain technology

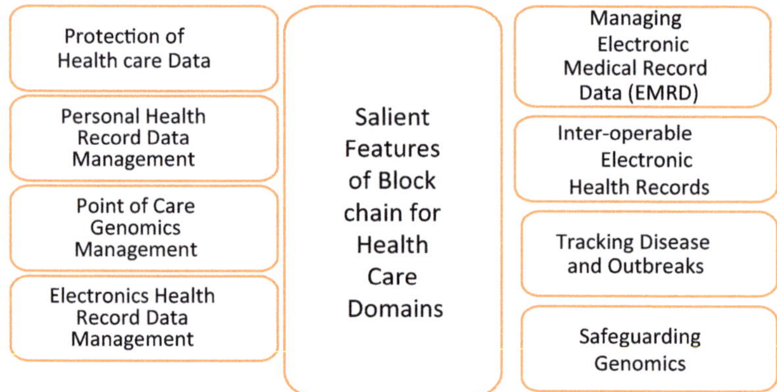

Fig. 2 Revolutionizing healthcare with blockchain technology: opportunities and applications

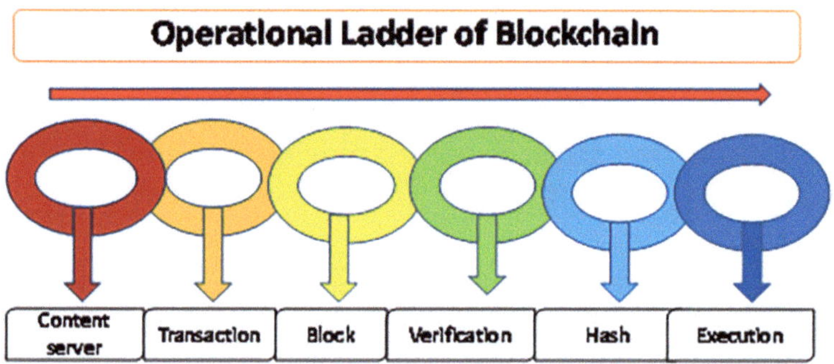

Fig. 3 Operational ladder of blockchains

[24]. By fully digitizing the components of the blockchain generation and using them in hygienic packaging, the entire process of drug manufacturing, traffic, transport direction, and speed can be tracked via IoT and blockchain, which reduces the risk of plan acquisitions, disruptions, and bottlenecks in ambulances, pharmacies, and various clinical centers. "Moreover, the blockchain technology offers a balance between visibility and privacy by using complex and secure algorithms to protect the confidentiality of clinical data while still being transparent. The decentralized nature of blockchain ensures that patients, doctors, and healthcare providers can access data quickly and securely. The patient-driven interoperability is also made possible by blockchain, giving patients control over their clinical data and access to prescriptions, which enhances patient empowerment and privacy [24]. Despite the challenges of achieving perfect control and scalability, blockchain solutions can help fix these errors and improve healthcare by allowing regulators to track fake drugs and ensure authorized transactions. Blockchain technology has been recognized by healthcare companies as a valuable tool, and its transformative potential for the healthcare market has been embraced by healthcare providers, patients, and the fitness environment. As a result, the use of blockchain in healthcare is expected

to expand in the coming years, providing easy access to all relevant data for doctors, patients, and pharmacists" by tracking drugs, increasing pricing options, and decentralization of data. In addition to the robust construction including synthetic intelligence, the clinical field is heavily dependent on the blockchain. "Blockchain technology can be applied in several ways to transform the healthcare industry. For instance, blockchain-based applications can be used to monitor the supply chain for accurate clinical refinement [25]. By using blockchain, it becomes possible to store extended statistics such as a patient's entire medical history, diagnosis, follow-up reports, diet information, and even measurements from smart sensors. Doctors can easily access all the necessary information to make accurate diagnoses and provide appropriate treatment recommendations. With all the data stored in a single blockchain device, it becomes less vulnerable to loss or tampering. Additionally, healthcare organizations can use blockchain to avoid the vulnerabilities of internal networks. For example, large agencies can establish differentiated control perimeters on encrypted blockchain databases and purchase external protection against potential attacks. By implementing a blockchain network, healthcare authorities can eliminate the need for data recovery efforts, such as rescue flights or the need to address issues such as laptop damage or hardware failure [26]."

6 Applications of Blockchain in Healthcare

As "Blockchain technology is a modern and rapidly developing era with innovative applications for efficient healthcare implementation. It enables seamless and efficient sharing and transmission of data across borders, leading to affordable medicines and advanced treatments for various diseases. The healthcare sector is poised to experience substantial development in the upcoming years owing to the blockchain era, which offers numerous opportunities in the logistics industry. Given that this field directly impacts people's quality of life, it is undoubtedly one of the first areas to embrace digital transformation and innovation." At the same time, the blockchain era is spreading, especially in the monetary area. It offers many important and unique elements: opportunities for health, technology, and logistics relationship between doctors and patients [27]. The feature of blockchain is to record all types of transactions in the decentralized ledger, except various health control structures. It is accurate and easy and successfully saves time, effort, and fees, saving management effort. The biggest problem encountered by a healthcare company is the disclosure of statistics and its use for malicious gadgets and various special interests that programs from that era can quickly solve. "Another important aspect of blockchain technology is its ability to provide real-time access to current and accurate data for customers and stakeholders. This has great potential for solving urgent business problems in the healthcare realm. By leveraging blockchain technology, we can connect healthcare providers, community services, and patients, ensuring everyone have access to the same material. The use of blockchain can bring significant benefits to the industry, including personal health records, clinical

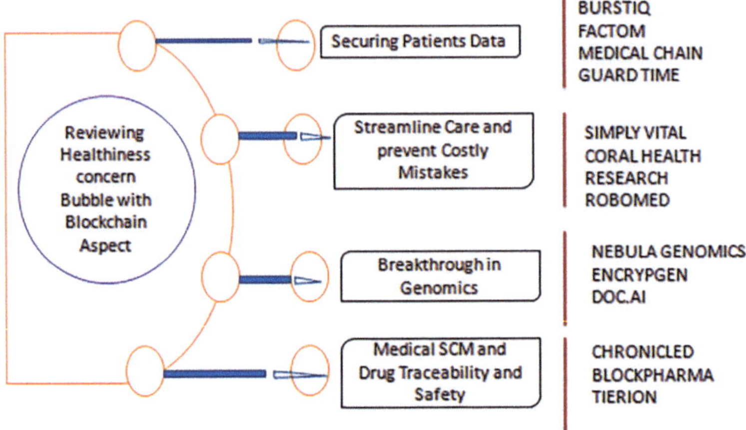

Fig. 4 Enablers of blockchain execution in healthcare services

technology, research, supply chain management, and drug integrity [28]. Figure 2 illustrates the various functions and key features of the blockchain philosophy in diverse health fields and allied domains."

7 Future Scope and Limitations

Interoperability states to the capacity of diverse systems to connect and interchange data with each other. Blockchain technology can support to improve interoperability in the healthcare sector through providing a protected and standardized way to share data between different providers, patients, and organizations [28].

We may anticipate seeing more initiatives in the future to create blockchain-based solutions that can aid in facilitating seamless data sharing between various healthcare groups. 2. Better patient data management. One of the most important aspects of healthcare is patient data management, and blockchain technology can assist to enhance how patient documents are reserved, retrieved, and shared. Patients will have more governance over their data, thanks to blockchain, while healthcare specialists will have access to more accurate and recent patient records. We may anticipate additional blockchain-based solutions that provide safe and effective patient data management in the future [29].

3. Improved supply chain management. Blockchain technology can also be applied to the healthcare industry's supply chain management. Using blockchain, it is feasible to trace the flow of medications and medical supplies from the producer to the consumer, promoting more accountability and transparency [28].

The era of blockchain has opened up new opportunities and challenges in the healthcare sector, which must be addressed. Although it is beneficial to use this advanced era for scientific purposes, the first level of blockchain packages needs to be expanded to explore and research the era. This task is not only for scientific institutions but also for regulatory authorities. It is time to boost the healthcare industry using blockchain technology, which is expected to grow in the future. The use of blockchain in healthcare is empowered by this technological innovation, which explains the implications and course of the recovery process wherever possible [29].

8   Conclusion

This work discussed the following:

1. The use of blockchain technology in healthcare to improve data clarity and immutability
2. The establishment of a new fitness service using blockchain technology
3. The development of a rights-based blockchain replica for storing EHR with enrolled patients
4. Ensuring eco-friendly equipment for doctors and patients through the use of this novel system
5. Potential improvement in public health by addressing issues related to slow, complicated, expensive frameworks prone to human error

In conclusion, the major contributions of this work are as follows:

1. Providing unprecedented opportunities for businesses and public sectors in ensuring health security which is a top priority for humanity
2. Improving data protection and privacy
3. Addressing issues related to slow, complicated, and expensive framework prone errors
4. Establishment of a new trending fitness service statistics with blockchain generation as platform
5. Contributing toward momentum advantage over utility next-generation blockchain at EHR

The healthcare sector has embraced modern blockchain applications due to the inherent encryption and decentralization, which offer enhanced security for patients' digital clinical records [30]. It also supports the monetization of healthcare information, improves interoperability between healthcare organizations, and helps fight counterfeit medicines. Several areas of healthcare can benefit from blockchain technology, including virtual contracts authorized by intelligent means, which represent one of the most valid blockchain use cases by keeping intermediaries out of the payment chain and reducing costs. The potential of blockchain in healthcare is vast, especially when applied to device tracking [31, 32].

Acknowledgments We are thankful to the management and Principal of C. Abdul Hakeem College (Autonomous), Melvisharam, Tamil Nadu, India, and to Dr. P. Srikanth, Systemics Cluster, University of Petroleum and Energy Studies, Dehradun, India, for their encouragement, providing the necessary facilities and support in carrying out the work.

Conflict of Interest The authors declare that there are no conflicts of interest.

References

1. Terzi, S., & Stamelos, I. (2018). Permissioned Blockchains and smart contracts into agile software processes. In *Proceedings of the international conference on software process improvement and capability determination, Tessaloniki, Greece, 9–10 October 2018* (pp. 355–362).
2. Wani, S., Imthiyas, M., Almohamedh, H., Alhamed, K. M., Almotairi, S., & Gulzar, Y. (2021). Distributed denial of service (Ddos) mitigation using blockchain - A comprehensive insight. *Symmetry, 13*, 227.
3. Al Mamun, A., Azam, S., & Gritti, C. (2022). Blockchain-based electronic health records management: A comprehensive review and future research direction. *IEEE Access, 10*, 5768–5789.
4. Reegu, F. A., Mohd, S., Hakami, Z., Reegu, K. K., & Alam, S. (2021). Towards trustworthiness of electronic health record system using blockchain. *Annals Of The Romanian Society For Cell Biology, 25*, 2425–2434.
5. Dar, A. A., Alam, M. Z., Ahmad, A., Reegu, F. A., & Rahin, S. A. (2022). Blockchain framework for secure COVID-19 pandemic data handling and protection. *Computational Intelligence and Neuroscience, 2022*, 7025485.
6. Hasselgren, A., Kralevska, K., Gligoroski, D., Pedersen, S. A., & Faxvaag, A. (2020). Blockchain in healthcare and health sciences - A scoping review. *International Journal of Medical Informatics, 134*, 104040.
7. Pournaghi, S. M., Bayat, M., & Farjami, Y. (2020). MedSBA: A novel and secure scheme to share medical data based on blockchain technology and attribute-based encryption. *Journal of Ambient Intelligence and Humanized Computing, 11*, 4613–4641.
8. Giungato, P., Rana, R., Tarabella, A., & Tricase, C. (2017). Current trends in sustainability of bitcoins and related blockchain technology. *Sustainability, 9*, 2214.
9. Allison, I. (2015). Bank of England: Central Banks looking at 'hybrid systems' using bitcoin's blockchain technology. *International Business Time*.
10. Goel, A., Sharma, A., Gupta, D., & Khanna, A. (2020, March 30). Immigration control and management system using blockchain. In *Proceedings of the international conference on innovative computing & communications (ICICC) 2020, Delhi, India, 21–23 February 2020*.
11. Beck, R., & Müller-Bloch, C. (2017). Blockchain as radical innovation: A framework for engaging with distributed ledgers. In *Proceedings of the 50th Hawaii international conference on system sciences, Waikoloa Village, HI, USA, 4–7 January 2017*.
12. Catalini, C., & Gans, J.S. (2017). *Some simple economics of the blockchain*. Rotman School of Management Working Paper No. 2874598; MIT Sloan Research Paper No. 5191-16.
13. Abu-elezz, I., Hassan, A., Nazeemudeen, A., Househ, M., & Abd-alrazaq, A. (2020). The benefits and threats of blockchain technology in healthcare: A scoping review. *International Journal of Medical Informatics, 142*, 104246.
14. Zheng, Z., Pan, J., & Cai, L. (2020). Lightweight blockchain consensus protocols for vehicular social networks. *IEEE Transactions on Vehicular Technology, 69*, 5736–5748.
15. Iqbal, M., & Matulevicius, R. (2021). Exploring sybil and double-spending risks in blockchain systems. *IEEE Access, 9*, 76153–76177.
16. Masseport, S., Lartigau, J., Darties, B., & Giroudeau, R. (2020). Proof of usage: User-centric consensus for data provision and exchange. *Annales des Telecommunications, 75*, 153–162.

17. Matulevicius, R., Iqbal, M., Ammar Elhadjamor, E., Ghannouchi, S. A., Bakhtina, M., & Ghannouchi, S. (2022). Ontological representation of healthcare application security using blockchain technology. *Informatica, 33*, 365–397.
18. Jabbar, S., Lloyd, H., Hammoudeh, M., Adebisi, B., & Raza, U. (2021). Blockchain-enabled supply chain: Analysis, challenges, and future directions. *Multimedia Systems, 27*, 787–806.
19. Azienda Ospedalieradella Provimcia di Pavia. Accoglienza Presa in Carico e Dimissione del Paziente (2016).
20. Pina, A. R. B., Torlà, C. B., Quintero, L. C., & Segura, J. A. (2017). Blockchain En Educación: Introducción Y Crítica Al Estado De La Cuestión. *Edutec. Revista Electrónica de Tecnología Educativa, 61*, a363.
21. Capps, M. (2016, November 8). ConsenSys Anticipates Moving Ujo Music Blockchain Rights Management Oering to Beta. *Bitcoin Magazine.*
22. Zhuang, Y., Sheets, L. R., Chen, Y. W., Shae, Z. Y., Tsai, J. J. P., & Shyu, C. R. (2020). A patient-centric health information exchange framework using blockchain technology. *IEEE Journal of Biomedical and Health Informatics, 24*, 2169–2176.
23. Parker, L. (2015, December 5). Provenance to restore consumer trust with the blockchain. Brave New Coin.
24. Bodkhe, U., Mehta, D., Tanwar, S., Bhattacharya, P., Singh, P. K., & Hong, W. C. (2020). A survey on decentralized consensus mechanisms for cyber physical systems. *IEEE Access, 8*, 54371–54401.
25. Reegu, F. A., Abas, H., Hakami, Z., Tiwari, S., Akmam, R., Muda, I., Almashqbeh, H. A., & Jain, R. (2022). Systematic assessment of the interoperability requirements and challenges of secure blockchain-based electronic health records. *Security and Communication Networks, 2022*, 1953723.
26. Sonkamble, R. G., Phansalkar, S. P., Potdar, V. M., & Bongale, A. M. (2021). Survey of interoperability in electronic health records management and proposed blockchain based framework: MyBlockEHR. *IEEE Access, 9*, 158367–158401.
27. Gill, E., Dykes, P. C., Rudin, R. S., Storm, M., McGrath, K., & Bates, D. W. (2020). Technology-facilitated care coordination in rural areas: What is needed? *International Journal of Medical Informatics, 137*, 104102.
28. Guimarães, T., Silva, H., Peixoto, H., & Santos, M. (2020). Modular blockchain implementation in intensive medicine. *Procedia Computer Science, 170*, 1059–1064.
29. Lahami, M., Maalej, A. J., Krichen, M., & Hammami, M. A. (2022). A comprehensive review of testing blockchain oriented software. *ENASE*, 355–362.
30. Taylor, P. J., Dargahi, T., Dehghantanha, A., Parizi, R. M., & Choo, K. K. R. (2020). A systematic literature review of blockchain cyber security. *Digital Communications and Networks, 6*, 147–156.
31. Reegu, F. A., Abas, H., Jabbari, A., Akmam, R., Uddin, M., Wu, C.-M., Chen, C.-L., & Khalaf, O. I. (2022). Interoperability requirements for blockchain-enabled electronic health records in healthcare: A systematic review and open research challenges. *Security and Communication Networks, 2022*, 9227343.
32. Song, H., Zhu, N., Xue, R., He, J., Zhang, K., & Wang, J. (2021). Proof-of-contribution consensus mechanism for blockchain and its application in intellectual property protection. *Information Processing and Management, 58*, 102507.

Music DApp on the Solana Blockchain Platform: Design, Development, and Analysis

Urmila Pilania, Manoj Kumar, Rohit Tanwar, Pulkit Upadhyay, and Priyanka Narayan

Abstract Decentralized applications (DApps) are the instructions or the software programs which run on a distributed system like a blockchain. Client-server architecture does not guarantee transparency and verification of data, but DApps are independent of client-server architecture. DApps through blockchain is immutable, maintain transparency and integrity, and verify the data as well as the user of the data. Such applications facilitate new directions for the building, deployment, and maintenance of software. In this paper, a DApp music player is designed and developed. The smart contract helps users to communicate data in blockchain programmatically. For the research work, Solana blockchain has been chosen because it is decentralized and provides very good transactions per second and low gas charges. The developed DApp also helps to safeguard the privacy of users and their access records. The research work contributes in maintaining the authenticity, integrity, and confidentiality.

Keywords Blockchain · Solana · Rust · Smart contracts · Transaction · Decentralized applications

1 Introduction

In today's world, people across all age groups access the Internet using various resources. A well-developed network environment is essential for connecting the entire globe to the Internet and ensuring the security and integrity of data transmission is equally important. With the widespread use of smart gadgets equipped with high-performance operating systems like iOS, Windows, Android, Tizen, and many

U. Pilania · M. Kumar · P. Upadhyay · P. Narayan
Computer Science Technology, Manav Rachna University, Faridabad, India

R. Tanwar (✉)
School of Computer Science, University of Petroleum and Energy Studies, Dehradun, India

© The Author(s), under exclusive license to Springer Nature Switzerland AG 2024
S. Pulipeti et al. (eds.), *Quantum and Blockchain-based Next Generation Sustainable Computing*, Contributions to Environmental Sciences & Innovative Business Technology, https://doi.org/10.1007/978-3-031-58068-0_6

more, ensuring data security and integrity in communication is becoming increasingly crucial. Blockchain technology provides a significant advantage in terms of data security and integrity. By using a distributed and decentralized chain of data shared between numerous nodes, blockchain efficiently maintains a record of transactions in digital form. Though blockchain has well-known usages in cryptocurrencies like Bitcoin, Ethereum, and Solana, it can also be used for ensuring the security of various forms of digital communication [1].

Solana is a high-performance, decentralized, permission less, open-source blockchain. It is a decentralized platform that efficiently attains very good transactions per second. It works on the concept of proof of history [2]. Proof of history is intended to maintain time between nodes on a decentralized system without all nodes having to communicate about it and come to an agreement. Solana can create smart contracts and can also communicate with smart contracts. With the help of smart contracts, we can create NFT markets, music players, lending applications, and many more DApps [3].

In Fig. 1, we can observe various notable features of the Solana blockchain technology. One of the critical features is proof of history, which ensures proper

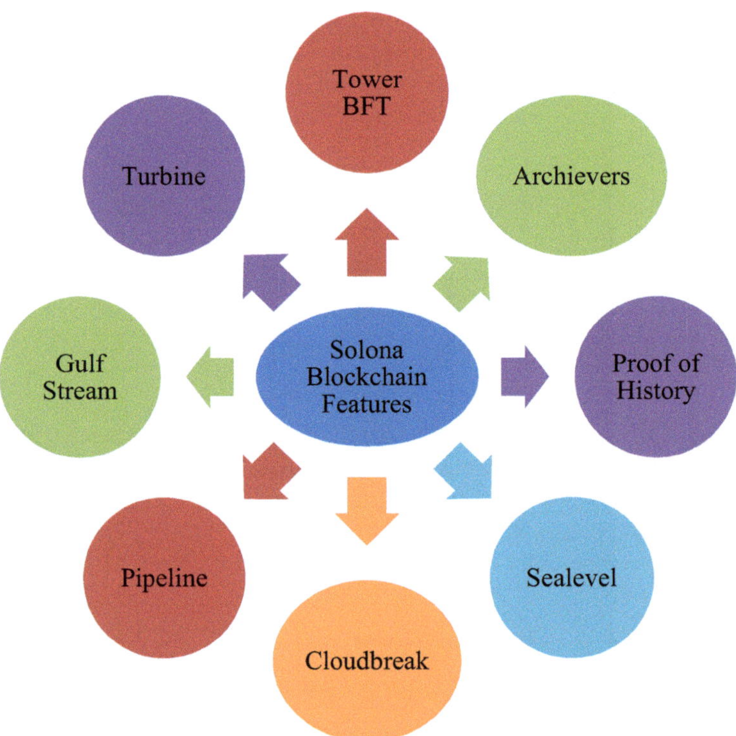

Fig. 1 Features of Solana blockchain [source: https://huddle.eurostarsoftwaretesting.com/8-key-features-of-solana-blockchain]

sequencing of cryptographic transactions and aids in their verification. Furthermore, Solana nodes function independently and use clocks, reducing transaction time. Another feature, Tower BFT, lowers transaction latency. Additionally, Turbine helps break data into parts for efficient transmission between nodes. Gulf streaming caches forward operations to the network's edge, decreasing confirmation time by ensuring transaction execution. Sea level is a parallelized transaction processing engine that improves runtime, while the pipeline assigns input data to hardware units for faster validation. To achieve a satisfactory level of scalability, Cloudbreak is another crucial feature. Finally, archivers play a pivotal role to ensure the data integrity in the Solana blockchain.

A comparison of the Solana blockchain with other famous blockchains is shown in Table 1. Comparison is done based on parameters such as transaction per second, gas fee, transaction delay, valuators, and the number of total transactions to date. From Table 1, it has been concluded that the Solana blockchain is gaining reputation day by day over other blockchains. In this proposed research work, we have designed and developed DApp for the music player. These applications work through the execution of smart contracts. The smart contracts are designed to leverage the Rust language, Anchor Framework. The contribution and motivations behind the proposed work are as follows:

- DApps help to safeguard the privacy of people.
- DApps also efficiently store their credentials.
- These apps help to maintain the authenticity, integrity, confidentiality, etc.

Paper is organized in total of seven sections. In Sect. 1, basic introduction of blockchain and DApp is provided. In Sect. 2, literature review of the existing techniques has been done, and some issues with the existing techniques are also listed after literature review. In Sect. 3, background and implementation process of the proposed work has been explained. In Sect. 4, DApp working is shown with the help of screenshots. In Sect. 5, comparison of the work is done with some existing work in the same field. In Sect. 6, some observations about the proposed work are listed. Finally the last section provides the conclusion and its future scope.

Table 1 Solana versus other blockchain [Solana vs. Polygon vs. Ethereum—the ultimate comparison—Blockchain Council (blockchain-council.org)]

Operations	Solana	Ethereum	Binance Smart Chain	Polkadot	Cardano	Tron
Transaction per second	65,000	15	10	1000	270	1000
Average fee per transaction	$0.0015	$15	$0.01	$1	$0.25	Free
Transaction latency	0.4 s	Approx 5 min	75 s	2 min	10 min	3 s
Number of valuators	702	11,000+	21	297	2376	27
Total transaction till date	15 billion	1.07 billion	227 billion	1.7 billion	5.9 million	1.7 billion

2 Literature Review

In IoT, the use of blockchain is limited due to the lower number of transactions per second. Researchers are working in the area of IoT with blockchain. Blockchain in IoT helps to maintain the authenticity and integrity of data. To overcome the low transaction rate, researchers tried the Solana blockchain so that the transaction rate can be improved. In the proposed research work, Solana blockchain using Anchor Framework has been proposed. The proposed work achieved 663 transactions per second. It has been concluded that using the Solana transaction rate could be improved [4]. However, centralized services lead to a lack of security and misuse of user data. After the introduction of blockchain, the concept of a decentralized web was introduced which efficiently provides peer-to-peer communication securely? Blockchains ensure users independently authenticate each transaction and enact their block to the existing web. In this work, authors [5] proposed CollaChain, which is decentralized and provides a high transaction rate. The user executes the smart contracts for validation of transactions and then adds his block to the web. The proposed blockchain allowed users to interact with each other securely using their mobile phones. It provided 4500 transactions for the decentralized Twitter application. It deployed 200 nodes in 10 countries covering 5 continents.

Moving ahead in paper [6], the author worked on whether blockchain is governed by the government or not. It is a common question among the people whether blockchain relates to government, in which blockchain is governed, and conducts that blockchains can advance scenarios for effective self-governance. The proposed work links the research by reconnoitering the suggestions of the Governing Knowledge Commons (GKC) context to examine the governance of blockchains. It has been concluded that blockchain trust on collectively managed skills to collect and accomplish dispersed information, explaining the expediency and originality of the GCK practices through an experimental studies of Bitcoin, and putting the basis for a research program employing the GKC method.

Training is the most important part these days because of the day-by-day evolution of new technologies. Managing training certificate for the participants is very challenging for institutions or companies. Managing a central database can act as a solution to the above challenge. But the central database also has some issues like security. In the proposed work [7], the authors designed CertificateChain for the storage and management of certificates. It is decentralized to avoid the issue of security also. It improves the transaction rate as well in terms of storage and retrieval of certificates. Decentralized finance (Defi) takes blockchain to the height of converting old financial products into transparent trustful protocols which do not require any intermediates. In this work [8], the authors pointed out the issues related to the Defi projects. In 2020, Defi reaches more than 15 billion $ in investment. Many decentralized blockchains had used the oracle database for the extraction of asset information from the outside world. Their prime and management standards were not known to the end users. If oracles were not selected properly, then funds growing of depositors may be in danger. It might cause a loss of millions of dollars invested by the stakeholders in the Defi projects.

For the last 5 years, blockchain has been widely accepted in business environments. Cryptocurrencies are widely accepted in many online businesses such as Bitcoin, Solana, or Ethereum in MasterCard expenditures. As blockchain is decentralized in nature, there is no option to degenerate completed transactions. This might cause a loss of cryptocurrency because of human error or an underprivileged user interface. In this work [9], authors designed a cryptocurrency wallet for creating accounts and tokens for the users. It results in enhanced user experience and avoids redundant errors. It carries many contracts which were deployed on the blockchain. After that, user registers to the account and token. So, there was no option to use an impervious address. In this way, the proposed decentralized blockchain application over-performs on centralized applications.

Some of the challenges still exist with the blockchain which are listed below [10, 11]:

- The number of transactions per second is less in many blockchains such as Ethereum, Bitcoin, Polkadot, Tron, etc.
- Centralized blockchains result in a lack of security and maybe in misuse of the Internet.
- As blockchain also provides the feature of decentralization, so in that case, there is no option to degenerate confirmed transactions.
- Blockchain might cause loss of cryptocurrency because of human error or underprivileged user interface.
- Smart contracts are not editable after their deployment on the blockchain even if it has some bug.
- In most blockchains, the transaction processing time makes smart contracts more deliberate resulting in low efficiency.
- As for smart contracts, there are limited standards so it is very difficult to maintain smart contracts.
- Some blockchains have high gas charges.
- Due to repeated calling of a particular function in a smart contract, sometimes the state variable does not update resulting in a serious issue.

3 Context and Execution of DApp for Music

In the beginning, web 1.0 served as a means for Internet users to access web content with the sole purpose of obtaining information through search. This version was essentially a read-only web. Later on, web 2.0 was introduced to allow users to access dynamic content that could respond to their queries. It provided users with the ability to create their own forms of content, such as blogs, podcasts, comments, and tags. Currently, web 3.0 is known as the semantic web, which is evolving into a set of tools that enable users to create, share, connect, and analyze content through search. Web 3.0 is decentralized and can be utilized for metaverse applications. It allows users to build blockchain decentralized applications that provide actual ownership. Figure 2 lists some of the features of web 3.0. Decentralized applications, which are interactive, are built using web 3.0, and smart contracts are utilized to interact with the decentralized applications and users.

Smart contracts are the instructions or agreements for the automation of certain transactions which are saved on the blockchain. These contracts execute when certain conditions are met with the consent of the buyer and seller. Smart contracts make transactions transparent, traceable, and immutable. As all the work is done digitally, no paperwork is required. All the information is stored in encrypted form so no security issues arise. Smart contracts have applications in industry, banking, telecommunication, NFTs, music, art, education, business, e-governance, and many more. Table 2 discusses some of the properties of smart contracts.

Blockchain is most widely used these days in every field of life. There are many types of blockchains such as Ethereum, Solana, Polygon, and many more. In this work, Solana blockchain is used to design the DApp. Solana can be written in C language or Rust language. Rust is the general-purpose programming language. It is difficult to write the program in Rust language compared to the solidity language. Programs written in Rust are scalable and attract the customer. It has high memory protection without using a trash collector. It is an open-source language. Its memory is managed internally. It is most widely used by the developer team. Its compiler works in parallel so errors are reported at the same time.

Solana uses a cryptographic hash function to verify the channel of period and coordinate diverse valuators on the system. The process is also known as proof of history. The first hash function will work as the input for the generation of the second hash as shown in Fig. 3. In this way, the chain is created, and it is also known as the proof of history. It is almost impossible to guess the hash function without its computation. Solana blockchains are collision-resistant because for the two

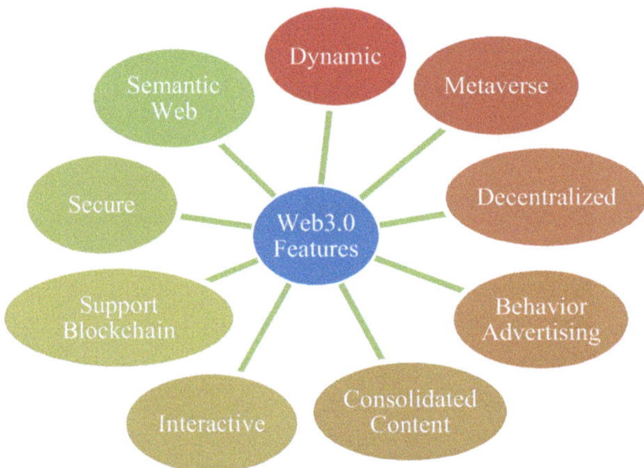

Fig. 2 Various features of web 3.0 [source: https://www.preface.ai/blog/trend/web-3-0/]

Table 2 Properties of smart contracts [12]

Parameter	Explanation	Advantages	Disadvantages
Transparent	It is public	The use of the application can be verified and validated	Need to be careful during implementation because of security purposes
Independent	It is designed and deployed in such a way that further interaction is not required with the agent who designed it	No intermediate is required for control and implementation	Operations once performed cannot be reversed
Unassailable	The contract code is not editable	The policy will not change in any case	The smart contract is not modifiable because of security concerns
Self-sufficient	It can coordinate with different functions for execution	After the deployment of a contract, no additional charges are required to be paid	The user of the smart contract needs to pay for accessibility

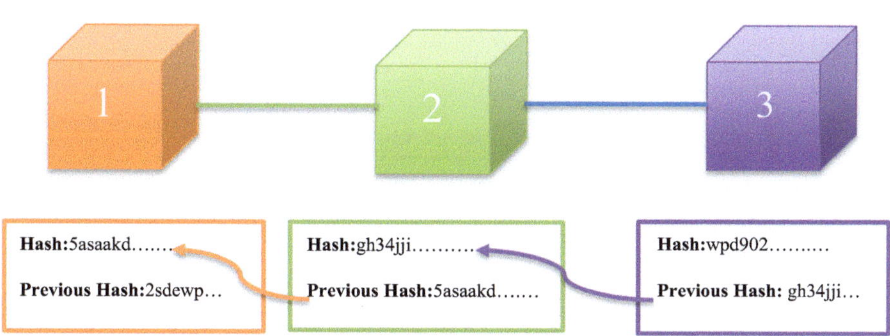

Fig. 3 Linking of blocks using hash generation in blockchain [source: https://www.guru99.com/blockchain-tutorial.html]

different inputs, it is impossible to generate the same hash function. The hash function can also be used to check the error.

$$h(x) = y \tag{1}$$

Here in Eq. 1, hash h produces hash as y for the input value x. It is practically impossible to generate the same hash for different inputs.

Figure 4 represents the DApp infrastructure for presenting results through smart contracts and the different layers of blockchain. For user interface, first step is to obtain the test token. Using the test token, gas charges are paid by the user. Gas fees are paid through the phantom wallet using airdrop. Now user can register by filling all the required fields and can make transactions. User registered detail can be used to communicate with user instead of blockchain address.

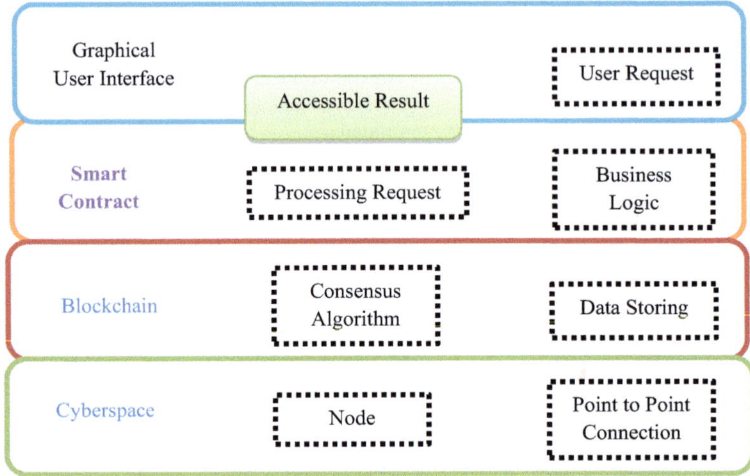

Fig. 4 DApp infrastructure [13]

```
PS D:\blockchain\spotify-clone-blockchain> npm run dev

> next dev

ready - started server on 0.0.0.0:3000, url: http://localhost:3000
event - compiled client and server successfully in 2.1s (540 modules)
wait  - compiling / (client and server)...
event - compiled client and server successfully in 544 ms (711 modules)
```

Fig. 5 Creation of local website

4 Working of DApp

To overcome some of the limitations in the existing work, a DApp for music is designed. DApps are decentralized to improve the security and integrity of data. There is no option to degenerate confirmed transactions because of the decentralized nature of the application. We designed Solana blockchain, so it supports a large number of transactions per unit of time compared to other existing blockchains. Solana blockchain has very low gas charges for transactions. For running the DApp, it first needs to build the website using **the nmp** command in the command prompt.

In Fig. 5, run the command **npm run dev** that builds the site from the local repo. It compiles the node template and begins with a local substrate-based Solana blockchain. It installs a front-end template to communicate with the local blockchain. It is then hosted locally on the web. The front-end template is used to acquiesce a smart contract and view the outcomes of operations. The URL (https://locall. host/3000/) is now open to make the gas payment. As the proposed work is based on the Solana blockchain, gas charges are very low here. As the creator of the DApp, we have the right to ignore these charges as well.

Here the gas charges are 0.1 Sol which is a very less amount compared to the other blockchains. The gas charges are also known as the transaction fee. Before the transaction can happen on the blockchain or be recorded, user has to pay a certain fee. Gas charges include network fees, wallet fees, and crypto exchange fees. On entering the abovementioned URL, user reaches the payment portal where payments can be done. Figure 6 is a locally hosted site's payment portal or one-time landing page. Here the user needs to pay 0.1 Sol to access the DApp.

When we click on pay, our phantom wallet extension pops up with a permission request as shown in Fig. 7. Phantom is the crypto wallet for the Solana blockchain. It helps to manage cryptocurrency and allows you to access DApps on Solana blockchain. Using this wallet, users can store, send, receive, and stake digital assets on the Solana blockchain. Now we make payment, and Phantom wallet pops up with a permission request to confirm the transaction as shown in Fig. 8.

Figure 9 represent Solfaucet, which is a site where the user can store the wallet public key and enter any desired amount that he needed to be airdropped in the wallet for testing purpose. These Sol tokens are only sent to dev net or test net servers. Dev net acts as a playground for users who want to work with Solana blockchain as an app developer, digital money holders, a validator, and users. Test net mainly focuses on network recital, constancy, scalability, and validator behavior. This Sol money is not delivered to the main net because the main net does not hold real value.

As the amount gets deducted from the user's dev net account, now the list is updated. Now the public key will be directly allowed from the next time whenever the user accesses again. In the Solana blockchain, the public key is the public address of the user. Now the user will land directly on the main site and will be able to play songs through DApp.

Now we can access the music file as shown in Fig. 10 with the help of smart contracts. Smart contracts are the encrypted line of code used to execute the agreement between two participants. Smart contracts allow safe and reliable transactions between users. Smart contracts in the Solana blockchain are stateless and got

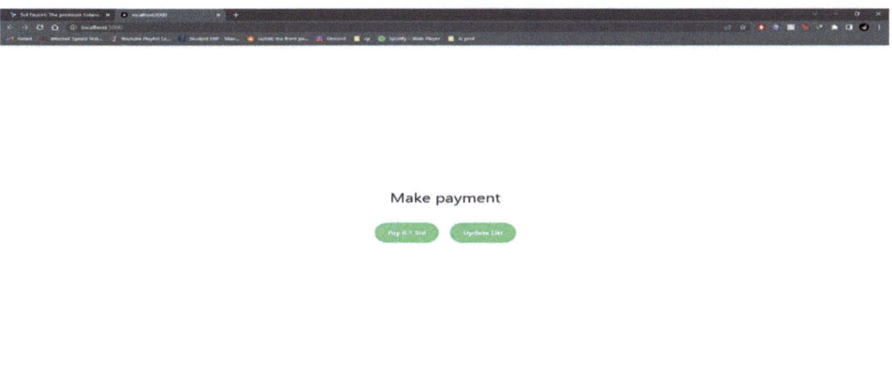

Fig. 6 Payment portal

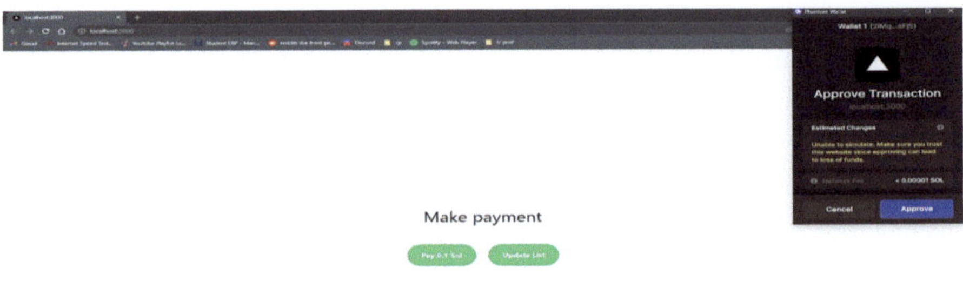

Fig. 7 Approval transaction request

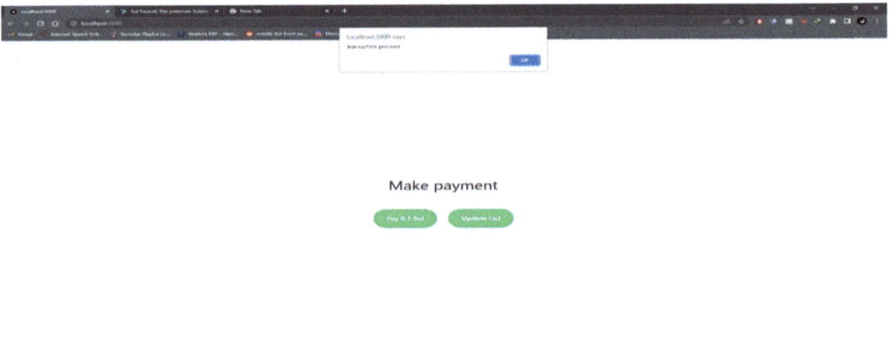

Fig. 8 Transaction processed

connected to external accounts once they have been installed. In this paper, the smart contracts are written in Rust language.

5 Comparison

The designed DApp is compared with some of the existing DApps in Table 3. We have compared work with papers of Moharrer and Nogoorani [14], Chang et al. [15], and Mendu et al. [1]. Comparison is done based on parameters autonomous, copyright, license validation, public blockchain, payload threshold, transaction cost, and latency. It has been observed from the comparison that the proposed work is able to handle all the mentioned properties.

There is no dependency on the government or any other organization to execute the transaction which means it is autonomous. Blockchain is able to determine

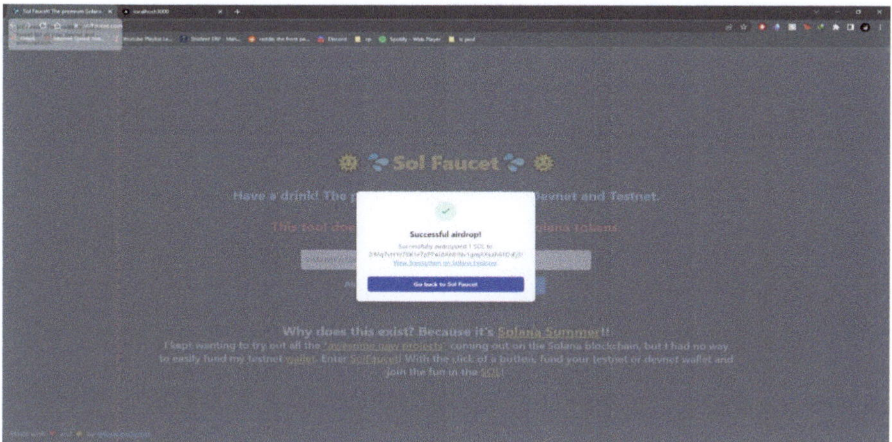

Fig. 9 Gas charges payment

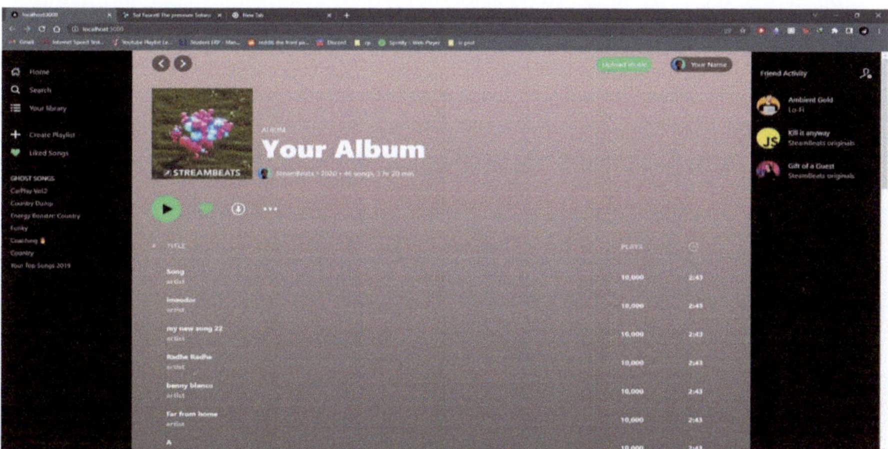

Fig. 10 DApp access for music

creator, proof of possession, and origin of the DApp. We have worked on Solana blockchain which is public and open-source. This is a locally hosted project and not registered yet. Payout threshold is the money that artists need to pay for uploading their music on the app; there is no payout threshold on our site which means it is a free service to upload. Solana blockchain has very low gas charges, and the creator of the application is eligible to do transactions without paying fee. Solana blockchain is very fast which allows a very less latency which is approximately equal to 0.4 s.

Table 3 Comparison of proposed work

Technique/ property	Proposed technique	Moharrer and Nogoorani [14]	Chang et al. [15]	Mendu et al. [1]
Autonomous	√	√	√	√
Reserve all rights	√	√	√	√
License validation	√	√	√	√
Public blockchain	√	√	√	√
Payout threshold	√	×	×	√
Transaction cost	√	×	×	×
Latency	√	×	×	×

6   Observations

A few observations for the proposed work are listed as follows:

Scalability. DApp is designed to be scalable, which means it is capable of handling a large number of users accessing the music DApp simultaneously without any significant reduction in performance. This feature is critical because it ensures that the DApp can meet the needs of a growing user base.

Traffic handling. DApp is able to handle the traffic more effectively and efficiently on the blockchain. Further, it provides the effective communication among the users without delays.

User authentication. DApp also provides the user authentication through verifying the credential before accessing the DApp which provides the authorization to individual user on this DApp platform.

Interoperability. DApp enables the users to interact with other users in the same or other blockchain network. Therefore, it improves the overall performance (functionality and usability) of DApp platform.

Security and transparency. DApp enables the trust among users and their transactions details are safer. Thus, it provides the transparency to all the transactions in the network and preserves in the blockchain in the one way hash format which ensures immutability and transparency.

Smart contract integration. DApp leverages the smart contracts which mitigates the privacy challenges of users which ensures the anonymity, self-execution, and enforceability.

Decentralized authority: DApp never depends on the third party authenticity; hence, it provides the more control over the user data.

Cost-effective. DApp leverages the proof of history consensus algorithm which is a cost-effective scheme, and it also processes the transactions more effectively and efficiently without requiring expensive computational devices.

User-friendly. DApp framework is user-friendly because the interface and processing allow the users to operate easily.

7 Conclusion and Future Scope

In this paper, we presented an independent DApp for music player that is supported by the Solana blockchain and smart contracts. The DApp did not depend on any mediators, alterations to the blockchain, or its transactions. It could be used by everybody, ranging from public users and developers to a great software retailer. This developed DApp does not have centralized management. This authorization is completed by distributing non-fungible tokens and bootstrap information that are safe to conduct to guard the applications. We have selected the Solana blockchain due to its acceptance, convenience of suitable development tools, a large number of transactions per unit of time, low gas charges, and high security.

In the future, we can design a DApp for securing the medical data of patients, as the medical history of a patient is very private and needs to be kept secured. This concept can also be used in chatting apps to safeguard the privacy of users and their conversations.

References

1. Mendu, M., Krishna, B., Mohmmad, S., Sharvani, Y., & Reddy, C. V. K. (2020, December). Secure deployment of decentralized cloud in blockchain environment using the inter-planetary file system. In *IOP conference series: Materials science and engineering* (Vol. 981, No. 2, p. 022037). IOP Publishing.
2. Pasdar, A., Dong, Z., & Lee, Y. C. (2021). Blockchain oracle design patterns. *arXiv preprint arXiv:2106.09349*.
3. Pilania, U., Upadhyay, P., Tanwar, R., & Kumar, M. (2023). A creative domain of blockchain application: NFTs. In *Mobile radio communications and 5G networks: Proceedings of third MRCN 2022* (pp. 397–406). Springer.
4. Duffy, F., Bendechache, M., & Tal, I. (2021, December). Can Solana's high throughput be an enabler for IoT? In *2021 IEEE 21st international conference on software quality, reliability and security companion (QRS-C)* (pp. 615–621). IEEE.
5. Tennakoon, D., Hua, Y., & Gramoli, V. (2022). CollaChain: A BFT collaborative middleware for decentralized applications. *arXiv preprint arXiv:2203.12323*.
6. Murtazashvili, I., Murtazashvili, J. B., Weiss, M. B., & Madison, M. J. (2022). Blockchain networks as knowledge commons. *International Journal of the Commons, 16*(1), 108–119.
7. Tellew, J., & Kuo, T. T. (2022). CertificateChain: Decentralized healthcare training certificate management system using blockchain and smart contracts. *JAMIA Open, 5*(1), ooac019.
8. Caldarelli, G., & Ellul, J. (2021). The blockchain oracle problem in decentralized finance—A multivocal approach. *Applied Sciences, 11*(16), 7572.
9. Bodziony, N., Jemioło, P., Kluza, K., & Ogiela, M. R. (2021). Blockchain-based address alias system. *Journal of Theoretical and Applied Electronic Commerce Research, 16*(5), 1280–1296.
10. Kushwaha, S. S., Joshi, S., Singh, D., Kaur, M., & Lee, H. N. (2022). *A systematic review of security vulnerabilities in Ethereum blockchain smart contract* (Vol. 10, pp. 6605–6621). IEEE Access.
11. Kushwaha, S. S., Joshi, S., Singh, D., Kaur, M., & Lee, H. N. (2022). Ethereum smart contract analysis tools: A systematic review. *IEEE Access, 10*, 57037.

12. Zhou, H., Milani Fard, A., & Makanju, A. (2022). The state of ethereum smart contracts security: Vulnerabilities, countermeasures, and tool support. *Journal of Cybersecurity and Privacy, 2*(2), 358–378.
13. Johnson, M., Jones, M., Shervey, M., Dudley, J. T., & Zimmerman, N. (2019). Building a secure biomedical data sharing decentralized app (DApp): A tutorial. *Journal of Medical Internet Research, 21*(10), e13601.
14. Moharrer, M. M., & Nogoorani, S. D. (2020, September). A decentralized app store using the blockchain technology. In *2020 17th international ISC conference on information security and cryptology (ISCISC)* (pp. 14–21). IEEE.
15. Chang, Y. W., Lin, K. P., & Shen, C. Y. (2019, March). Blockchain technology for e-marketplace. In *2019 IEEE international conference on pervasive computing and communications workshops (PerCom workshops)* (pp. 429–430). IEEE.

The Role of Blockchain in Healthcare

Radhika Sreedharan

Abstract The healthcare industry generates a lot of data, and much of that data is scattered around. Patients go to general practitioners, specialists, hospitals, outpatient care clinics, and other locations for health needs. All of these visits generate data. Pronouncements are parallel which need that healthcare issuers enable patients to access their every digital data. Google is employing a verifiable data inspection, a system based on ledge which can corroborate data evidences cryptographically. The specific objective of this paper is to describe the function of blockchain in healthcare domain.

Healthcare is characterized as a conventional profession which is substantially inflexible for measuring because of certainty of variation and compliant to innovatory implementations. Situations in healthcare like secrecy, level of protection, and protection of information are illustrating awareness in present days globally. Blockchain techniques are growingly recognized as a mechanism for addressing confirmed information division concerns. It can enhance instantaneous healthcare procedures, for instance, in enhancing health service distribution and standard of keeping up protection. A current outline of Market Research Future (MRFR) proves that blockchain technique in healthcare is presumed to produce more than 42 million in value and attain a consolidated yearly expansion speed of 71.8% by 2023. Such capable increase is directed by taking over blockchain peculiarities of distributed log technology having substantial clarity, enhanced protection and secrecy, enlarged identifiability, improved effectiveness, and decreased costs.

The basis of peer-to-peer network, a boon to utilize blockchain is it revises instantaneously and does not leave scope for representatives and related costs. Being resilient to reconstructions, blockchain issues a clear surrounding in which health professionals and patients can access records apparently and in the absence of additional costs. It also grows protection of the system by decreasing the possibility of missing evidences and inaccuracies.

R. Sreedharan (✉)
Presidency University, Bangalore, India
e-mail: radhika.sreedharan@presidencyuniversity.in

© The Author(s), under exclusive license to Springer Nature Switzerland AG 2024
S. Pulipeti et al. (eds.), *Quantum and Blockchain-based Next Generation Sustainable Computing*, Contributions to Environmental Sciences & Innovative Business Technology, https://doi.org/10.1007/978-3-031-58068-0_7

The purpose to apply blockchain in healthcare emerges of the requirement for protection and intercommunication in healthcare. By the development of IoT and plentifulness of health gadgets and mobile healthcare applications, a significant quantity of therapeutical data is transcribed and transmitted daily. This data traffic requires direction concerning secrecy and protection. Blockchain technology can issue a solution which not only assists for protecting, registering, and disclosing of medical evidences but also ensures secrecy of every patient's data by issuing the patients their therapeutic data possession. Apart from the benefits of blockchain for healthcare supervision, its ultimatum has to be met first. Citizens can participate in programs related to health study utilizing blockchain technique. Additionally, finer exploration and divided data on widespread security will improve treatment for various groups. A centralized record is utilized for managing the whole healthcare system and organizations.

Keywords Blockchain · IoT · Healthcare · Data traffic · Computerized health evidence

1 Blockchain

The blockchain has got a lot of cryptographically corroborated data. Databases that are deficient in cryptographic integrity assure which are needed to work in combating surroundings, which is one of the causes for acceptance of blockchain. It emerges that blockchains which guarantee security and privacy gain acceptance more quickly. The instructional group has to perform jointly, at present a lot more than earlier, for sorting frauds of blockchain from the one which issue extremely useful offering. COVID-19 got an overall impact which is exceptional. Due to its utmost communicability, this infection has affected and destroyed a considerable amount of the general public. COVID-19 has an importantly deleterious influence on the general frugality due to serious justifying estimates; for instance, the strategy utilized by governments worldwide is a lockdown. The indications of COVID-19 differ in strength from individual to individual, and the typical indications are exhaustion, wheeze, fever, and gasp shortening, which are similar to indications of the influenza virus. But not everybody who is troubled with the sickness suffers indications. Symptoms of illness will not be shown by silent carriers or silent spreaders, but they carry them and transmit them to others. Moreover, the disease has a period of 2 weeks. As an affected individual may affect individual despite not going through any indications at this time, detailed and diversified checking is needed for an effective COVID-19 reaction.

Blockchain technology has the potentiality for transforming the profession of healthcare by establishing current models to manage and handle healthcare records securely and also maintain regulatory compliance. Blockchain technology is used for keeping track of public health data, in relation to these detections, mostly at the time of contagious disease epidemics such as COVID-19. At the time of virus epidemics, a blockchain can be utilized for reporting energetic and recuperated patients, permitting district and federated functionaries for preparing medical prerequisites and fast care, decreasing the spread of epidemics. But research detections reveal that at the time of examining data applicable to the COVID-19 catastrophe as per Razzaq et al [9], an important ultimatum in management of data which the medical performers and researchers faced was a deficiency of assimilation of date, genuineness, and origin point. The volume of data divided among health centers, government corporations, and social/nonpublic collaborations was undetermined or hard to propagate, as per the World Health Organization (WHO) and Centers for Disease Control (CDC) and Prevention.

Administrations evolve and execute acknowledgement planning on the basis of data related with diseases congregated by local units, as revealed at the time of COVID-19 disease. Medical detections through hospitals and other specific institutions are the origin of this data. But, the subsistence of many mediators in this procedure generates disclosing delays, as a result restricting capability of hospital and testing centers for responding fast when diseases are disclosed. Further, a layered disclosing form can cause variabilities, negatively influencing overall reaction plans and their capability for reducing illness.

1.1 Literature Review

The motivation of this research is for addressing the problems of punctual and accurate disclosing of COVID-19 viruses for supporting attempts of illness reactions. The blockchain technology has been accompanied in a current period of application establishment, one that is thoroughly dependent on the data structure of Bitcoin app being executed accurately. There are innumerable actual implementations in a variation of domains, which include the IoT, e-government, and management of electronic data that are currently produced. Due to self-cryptographic validation structure of peer-to-peer network and the general obtainability of a decentralized logs of proceedings evidences, these applications utilize blockchain technology and smart apps.

Blockchain was initially launched as a procedure for powering Bitcoin as per Tandon et al [7] but has recently developed relevance of being mentioned as a fundamental technique for numerous distributed implementations. Blockchain is recommended as a practical technique to manage sensitive data, mostly among the areas of healthcare, therapeutic analysis, and coverage. Healthcare can be analyzed as a means which incorporates three main elements: (a) fundamental issuers of therapeutic care facilities, like physicians, nursing assistants, supervisors of health centers, and technicians; (b) crucial tasks which are linked with medical operations, like medical exploration and health coverage; and (c) patients or public who are recipients of medical and health-familiarized services.

According to Talesh and McCoy and Perlis, in the area of healthcare, security and privacy violations are seemingly growing each year, with greater than 300 violations disclosed in 2017 and 37 million therapeutic evidences influenced from 2010 to 2017. As per Meinert et al., the growing digital conversion of healthcare has additionally assisted the acceptance of matters assisted with protected storage, possession, dividing of sufferers' unique health evidences, and integrated therapeutic information. As per Rupasinghe et al., blockchain is proposed as measure for solving crucial problems faced by healthcare, like protected dividing of medical evidences and agreement with regulations of secrecy of data.

As per Ago et al., Angraal et al., and Holbi et al., earlier investigation has produced restricted efforts for encapsulating remaining awareness in a holistic manner by using systematized report explorations. For example, Hölbl et al. utilized scientometric approaches for presenting a review of blockchain components and research directions which pertain to the implementation of blockchain in healthcare. In 2017, Angraal, Krumholz, and Schulz elaborated the numerous principles which are established for deploying blockchain in healthcare. In 2019, Agbo et al. described dissimilar illustrations of the selection of blockchain technique in healthcare, the issues encountered, and feasible resolutions. O'Donoghue et al. described particular substitutions and preferences of schematic implemented by investigators in numerous situations in which blockchain technique was utilized in 2019. In the same year, Jaoude and Saade organized work which pertains to implementations of blockchain over many industries and widely described the dissimilar utilization conditions for this technique. Currently, Hasselgren et al. [1] understood 39 studies for presenting synopsis facts on approved principles and attacked fields in which blockchain is utilized for enhancing healthcare.

As per Hasselgren et al. [1], in the systematized report exploration provided to the existent framework of awareness, their priority has fundamentally been on incorporating or outlining directions in 2020 and blockchain implementation fields. But, because of the scope and dissimilarity of preceding investigation on blockchain, investigators can gain through a concentrated description on the consequences for its acceptance and particular issues and fields for enhancement for proceeding the area. As per Agbo et al. and Ozdagoglu et al., studies on the basis of review can help to meet these requirements by incorporating effective awareness and interpreting important fields which require important academic attentiveness.

We direct this requirement by managing systematized reports exploration on the utilization of blockchain in healthcare as per Kitchenham et al. As per Aznoli and Navimipur in 2017, systematized report exploration can issue a valued encapsulation of recent awareness in an area of exploration and permit for identifying substantial awareness gaps and, as a consequence, approaches for upcoming exploration as per Gopalakrishnan and Ganeshkumar in 2013. This work provides to the recent reports on blockchain in healthcare by including to earlier systematized report exploration in two methods. Firstly, it issues a restrictedly structured, innovatory categorization of earlier studies with concerning areas of applications, restrictions, and directions. Second, on the basis of discoveries of the systematized reports exploration, we suggest an incorporating foundation for detailing potential themes which need academic attentiveness for advancing the recent framework of awareness. This subscription is produced by directing four experimentation queries (EQ):

EQ1. What is the inventory research outline for blockchain implementations in field of healthcare?

EQ2. What are the major fields of healthcare in which blockchain is utilized?

EQ3. What are the developing restrictions and ultimatum that the study postulates for this field of research?

EQ4. What are the hereafter approaches in healthcare which may gain through blockchain application?

Blockchain is a decentralized global log record which is serviced through the network of corroborated nodes and gathers unchangeable data blocks which can be allocated protectively in the absence of mediator involvement. As per Mendling et al., data are maintained and set down using secret writing marks and general agreement procedures which are approved as essential facilitators of its utilization. This capability for data maintenance is an important cause which has directed blockchain utilization in healthcare as per Kuo et al., where an important portion of data is put through to substantial interchange and division according to Meinert et al.

The development of blockchain technique and its implementation in various circumstances have happened in numerous stages. The initial stage of blockchain development has been associated with cryptocurrency, and the subsequent was applicable to the implementation of smart agreements in fields like actual ground and money matters as per Swan and Agbo et al. The next stage of development was concentrated on the blockchain implementations in nonfinancial disciplines like authority, healthcare, and tradition. In addition, managed through advanced scientific characteristics like data rigidity as per Yli-Huumo et al. in 2016, blockchain is currently reviewed to be in its final phase of development surrounded by essence of artificial intelligence (AI) as per Angelis and da Silva in 2019. Blockchain's maintained variation in its extent of implementations can be allocated to its potentiality to create distributed and suspicious proceeding surroundings.

The main possibility for blockchain technology is the healthcare industry as per Kuo et al., Alla et al. [2], and Cios et al., because blockchain maintains potentiality for addressing crucial concerns, like self-activating assertion evidence as per Angraal et al. and worldwide supervision of health as per Mettler. This technique can permit sufferers to possess data and select with whom it is divided, owing to address existent matters regarding ownership and dividing of data. Parallelly, it permits data evidences to be consolidated, revised, interchanged in a secured manner and retrieved punctually by suitable officials utilizing the concurrence procedures [2]. This is a vital benefit sustained by the utilization of blockchain technology subsequent to healthcare scope due to recent implementations of the need for data to be saved with intermediaries. At last, blockchain can possibly give rise to clarity to procedures of data administration and maintenance at the same time decreasing the possibilities of data misusing due to feasible human inaccuracy [2]. In spite of the explicit implications of blockchain's influence on social and business change, there appears discussion on its widespread benefits and obtained advantages in contrast to earlier developed predictions. A current description indicates that though associations will take up important speculations in executing technologies based on blockchain hereafter, they will likely accept a carefully empirical procedure due to a widespread assumption that the advantages may be over-increased. It can be said that this technique has to satisfy its recommended predictions someday, a truth that may be allotted to definite problems to the worldwide execution of this technique, mainly with reference to controlling obstacles. Other significant issue in publicizing the establishment of blockchain is the unawareness of the general and distinctive users, like patients or doctors, in the manner this technique performs, its technological characteristics [2], or its advantages for supervision of data. Iansiti and Lakhani recommend that because of general, operational, and execution hurdles, like protection or governance, important moment might be needed for blockchain for generating the predicted positions of commercial process variation. According to Swan and Alla et al., this can be furthermore being made up of widespread unpredictability about blockchain's utilization regarding legalized agreement and administration prescriptions. Recent investigation is concentrated on helping the functioning development of blockchain and speeding ups its extensiveness by directing these ultimatum and hurdles.

1.2 Blockchain Technology Is a Big Deal

A blockchain is a distributed collection of records which is divided within computer network nodes. A blockchain performs like a database, storing information in a computerized form. The word "blockchain" emerges through the certainty that "blocks" or transactions (records) are established and connected to the previous "block" for building an extended sequence.

Blockchain can be thought as a huge Google doc file with an important dissimilarity: anybody can access and add to it, but will not be able to revise it. Blockchain and healthcare arrange cooperatively by incorporating every collaborator's data in one position for them for viewing and performing with; variations done by one person in the data are obtainable for all in the system.

The features of the blockchain in healthcare environment as per Kanika et al [11], integrate for addressing various of recent issues in the healthcare industry. Medical blockchain is a game-varying discovery and a big contract as it resolves digital problems such as the following:

Transparency: Due to distributed form of blockchain, every transaction can be clearly viewed by using a unique node or discoverers of blockchain. It permits anybody for viewing transactions in real time.

Security: Data is kept over every system in blockchain instead of centralized database, which enhances protection and makes it hack-proof.

Cost-efficient: Blockchain technology is one of the most innovative techniques. By permitting peer-to-peer transactions, blockchain in healthcare removes third parties and even banks, decreasing costs of transaction.

Immutableness: When data is moved into the blockchain, nobody can vary them. It assists people to demonstrate that their data is genuine and unaltered.

Systems' integration: Blockchain in healthcare issues a distributed computer network and Internet which spreads topographical borders. Thus, when one health organization asks for patient's information from another, they need not wait longer for the data for traveling several miles.

Innovation: Blockchain issues a tremendous principle for a recent and advanced company models for growing and competing in opposition to established businesses.

1.3   Categorization of Blockchain

Blockchain is categorized, as per Xi et al [6], as local chain, nonpublic chain, and association chain in relation to the involvement practice. A local chain is totally social and obtainable to anybody. As the data present in chain is not able to be changed, local chains are estimated to be distributed completely. The association series is only restricted to approved representatives to involve, and the write and read authorizations and collaboration grouping authorizations on the blockchain are developed as per the regulations of the association. The nonpublic sequence is only utilized in nonpublic associations, and the write and read authorizations on the blockchain and the authorizations to involve in bookkeeping are developed as per the regulations of the nonpublic corporations. Involving nodes are hardly any and strictly limited.

1.4   The Following Functions Are Performed Well by the Blockchain

1. Tracing/recording: registering information and data in fixed and clear manner, in which no party has unsymmetrical power around the data.
2. Data authorization/transmission: facilitating data transmission among multiple parties, for creating a standard origin of accuracy.
3. Identity/authentication: undertaking identities and authorizations to authenticate or verify, which includes capability for verifying identity attributes in the absence of disclosing confidential information.
4. Settlements: agreement of revenue by registering mobility of products/returns or utilization of tasks/properties.
5. Proceedings: facilitating (actual) remittance and proceedings
6. Interchange of tokens: virtual currency/tokens having built-in value dealt among many intermediaries. Virtual currencies can also be fixed to precept currencies, with identical values held in charge accounts.

1.5   Business Advantages Delivered by the Abilities

- Security. Blockchain is corroborated with the help of a consensus system and stored over a lot of nodes, creating DDoS cyberattacks and tinkering with evidences very tough.
- Cost-effectiveness. Middlemen who take a cut of transactions can be removed as the agreement methods build faith with the help of clarity.
- Traceability. An invariable record of every transaction can decrease fraud and secure opposing to liability.
- Commercial operations process rate. Spontaneous smart agreements can decrease transaction time as the procedure does not need hand-operated error.
- Token value. Computerized properties can hold virtual and real-world value, such as when a virtual token is utilized for a loyalty points program.
- Sensitivity. Associations among organizations happen by not disclosing delicate information, like personal medical evidences.
- Equal and neutral. Organization or personnel cannot possess the blockchain, reassuring reliability and durability of the system, for instance. When any establishing party leaves, the system will carry on to function in their absence.

Specifying the big ultimatum healthcare systems are encountering around digitizing and disclosing therapeutic evidences, and tracing recommended medicine and other therapeutic products in the supply chain and dispatch, it is not unexpected that many are attempting to enhance methods in healthcare by utilizing blockchain technique.

In the meantime, utilizing many blockchain solutions for healthcare does not require profound knowledge from the origin source with the technology anymore, as a lot of blockchain-based resolutions are issued now like any other software-as-a-service.

2   Healthcare

The healthcare industry generates a lot of data, and much of that data is scattered around. Patients go to general practitioners, specialists, hospitals, outpatient care clinics, and other locations for health needs. All of these visits generate data. Procedures are coming up which needs that healthcare issuers allow patients for accessing their every digital data. Google is working on verifiable data audit, a ledger-based system which can cryptographically confirm data records.

Healthcare is characterized as a standard industry which is highly intractable for measuring because of the truth of variation and being resistive to innovatory implementations. Concerns in healthcare (e.g., secrecy, standard of care, protection of information) are illustrating awareness in current period globally. Blockchain technologies are progressively identified as a mechanism for addressing genuine data decentralization concerns. It can enhance instantaneous healthcare implementations, for illustration, in enhancing health assistance distribution and standard of protection maintenance. A proficient establishment is operated by taking over blockchain properties of distributed log technique having more clarity, enhanced protection and secrecy, enlarged trackability, improved productivity, and decreased costs.

On the basis of peer-to-peer networks Prokofieva et al. [3], an advantage to utilize blockchain is it improves with actual time and does not leave scope for negotiators and linked prices. Blockchain is insusceptible to updating and issues a clear surrounding in which healthcare workers as well as patients can retrieve evidences apparently and in the absence of appended costs. It has also grown protection of the system by decreasing the probability of missing evidences and inaccuracies.

The plan to apply blockchain in healthcare emerges out of the protection and coordination requirement in healthcare. A substantial quantity of medical data is recorded and transmitted daily by the development of IoT and plentifulness of health gadgets and mobile healthcare applications. This data traffic requires supervision with respect to secrecy and protection. Blockchain technology can issue a resolution which not only assists to protect registering and disclosing of therapeutic evidences but also ensures secrecy of every patient's data by issuing the patients of their therapeutic information possession. In spite of the benefits of blockchain for supervision of healthcare, its issues ought to be met foremost.

Blockchain is distributed ledgers, as per Sharma et al [14] which are mutually run by users called miners present in the network. This permits blockchain from any of the tendencies of governments or organizations and permits for the formation of fixed data evidences that are permitted to utilize over the world. Because of their distributed characteristic, blockchain has a huge quantity of computational power backing them up which makes it practically unfeasible for them getting hacked. Blockchain also permits for users on the Internet to be protectively recognized utilizing their cryptographic private keys. These two features of blockchain can be consolidated for creating data management systems which are fully flexible to hacks as they don't encounter from a single point of faultiness. As a matter of fact, ID supervision startup civic utilizes blockchain for this motive—they upload a computerized corroborated replica of a user's ID documents such as passports or driving licenses to their blockchain. Subsequently, when an entity requests access to that document, the user receives a trigger on the phone through which they can select to permit or deny the request. The same model can be used in healthcare evidences as well so that people are not left in the dark about their confidential information being revealed like in the case of the Equifax hack. This also has the additional benefit to simplify data origin—users can select for sharing medical data with researchers unrecognized so that their outcomes are utilized for developing better treatments in the subsequent time. There are no representatives in these proceedings, and healthcare researchers ensure that integrity of data is not adjusted.

3   Blockchain for Healthcare

Blockchain improves healthcare organizations for issuing sufficient care for patient and high-standard health prerequisites. Health information exchange is a prolonged and unchanging procedure that causes high health industry costs, organized fast utilizing this technology. Residents can participate programs of health study utilizing blockchain technology. Additionally, superior research and divided data on general welfare enhance treatment for various groups. A consolidated record is utilized for managing the whole healthcare system and corporations.

Blockchain has a broad scope of uses and tasks in case of healthcare. The log technique assists healthcare investigators expose genetical code by promoting protected transmission of patient medical evidences, supervising the drug supply chain. Security of healthcare data, supervision of numerous genetic, electronic data management, medical evidences, coordination, computerized tracking and problems breakout, etc., are certainly the technically built and magnificent characteristics working to establish and implement Blockchain technology. The entire computerized features of blockchain technique and its utilization in healthcare-linked implementations are the important causes for its acquisition.

Blockchain improves the protection of patients' electronic medical evidences, advances the health information's monetization, enhances interactivity between healthcare organizations, and assists fake fighting medicines. Dissimilar healthcare areas may vary with blockchain technology; fields such as healthcare and computerized contracts permitted by intelligent agreements comprise one of blockchain's most crucial applications. Intelligent agreements will reduce costs when intermediaries are removed through payment chain. The blockchain capability in healthcare relies importantly on the approval of related progressive techniques in the environment. It incorporates following of system, healthcare provision, tracing of drugs, and clinical examinations.

The healthcare area is improving its perspective and enlarging its utilization to place recent technologies like Internet of Things (IoTs). Remote Patient Monitoring (RPM), a healthcare application, is used to collect important patient indications from patients outside of the physical clinic location and keep track of them. As per Haleem et al [4], RPM incorporates various kinds of sensing elements that can be implanted or worn. These sensing elements transmit information using wireless communication till localized basis station, which transmits the information to an intermediate supervising station that alerts doctors to interrupt efficiently using patient's condition. All this data is saved in EHR (electronic health records).

Consequently, digital fitness evidences are a computerized pattern procedure of a patient's fitness data produced and maintained at the time of survival of a patient. It is normally fetched between numerous health centers, healthcare clinics, and care issuers. Digital fitness evidences issue significant and very sensitive individualized data for healthcare estimate and supervision. Hence, disclosing healthcare information needs to be cautiously supervised for preserving the data protection and build up healthcare assistance standard.

The issues encountered by computerized healthcare evidences and RPM when approved by healthcare systems are utilizing localized records for storing and managing access evidences of their patients' therapeutical information. But, the utilization of localized records sets up a lot of issues regarding data and information safeguard, which includes data secrecy, truthfulness, and interactivity. For overcoming these protection problems, blockchain (BC) technology in this circumstance will issue information evidence and trustworthiness and will also assist to spread data inside the system and between the numerous therapeutic requisites. The characteristics which the blockchain brings influence the cost, data standard, and significance of issuing healthcare within a clear, distributed network removing the requirement for an intermediary or any unified authorization. Remaining benefits issued by BC are the dependable authentic proceedings and the effective information access for the corroborated approved parts of that BC system. The BC ecological system can also eradicate issues of permitting patients for sharing information, which is the major obstacle for creating a frame to exchange health information. Recent health data interchange measures form an agreement with the taking one's place or lay hold of the uncomplicated eccentric model; however, they normally have to include various conditions to disclose extremely confidential information associated to the patients' medical background, which includes every essential controlling clinical data that are under control with the supervision which a specific issuer gives to an individual like population, issues of progress outcomes, medications, significant indications, medical background, reports of immunization, laboratory data, and reports of radiology.

Furthermore, technology of BC has the capability to use time stamping for validating variations to a record that is appropriate to handle computerized healthcare evidence data, especially when more than one user has the permission for record editing.

Apart from that, blockchain issues a probable succeeding policy to share data in assistance of a cooperation and certainty system that would permit revealed clinical healthcare resolution-producing in precise drug and telehealth.

Lastly, patient healthcare information could be either gathered through wearable and comfortable sensing elements or found by users manually. If this data is saved in a centralized record, it can produce a reason beyond data concurrence because of variance of healthcare issuers. Subsequently, utilizing a distributed database to save healthcare information permits the synchronized data obtainability.

Blockchain as a trustworthy, protected, and clear decentralized log technique or distributed ledger technology (DLT) has a considerable capacity to resolve the earlier recommended issues and thus can be utilized for the supervision of patient' computerized healthcare evidence and the protection of the system which monitors patients remotely. HealthBlock on the basis of framework influences both BC and IoT techniques for assuring protected healthcare supervision system which includes RPM and computerized health evidence disclosing. HealthBlock permits patients for managing their healthcare data on their own protectively. The essential subscriptions are outlined as given below:

- The blockchain technology utilization for ensuring supervision of computerized healthcare evidence, control of permission, integrity of data, scrutiny, and interactivity.
- The usage of distributed records to reduce vulnerabilities in centralized storage and for assuring adaptability of system.
- The suggested architecture was established utilizing Hyperledger Fabric and Hyperledger Composer as per Zaabar et al [5] for assuring protection needs and for increasing its performance.
- The suggested architecture's task is assessed utilizing Hyperledger Caliper.

4 How Startups of Blockchain Transform the Healthcare Area

A group of development interpreters explored 300 blockchain newly created corporations and made a broad-based synopsis of blockchain uses in the healthcare industry.

4.1 Supervision of Patient Data

Regarding supervision of data, the healthcare industry encounters an abundance of problems: through ineffective therapeutic awareness transmission to individualized medical attention, blockchain assists professionals to handle these scenarios in a better way. The problem which is frequently aspersed is there is possession of complete data by a patient. Therapeutic evidences can be utilized and/or shared in the absence of clear permission of a patient. Blockchain directs this problem by collecting patient information and saving it in an interior record of corporation. Smart agreements and their appropriate regulations control the access of patient data. A patient will be able to share their medical evidences in this manner. The data can be unidentified for the purposes of research or made fully clear, relying on patient's requirements.

Furthermore, by appending main regulations to these smart agreements, a patient will be able to reveal their health data and figures of sensing elements (e.g., through their smartwatch) with their doctor for making subsequent treatment much more efficient and individualized for achieving finer outcomes.

4.2   Clinical Resolution

Clinical examinations are a crucial part of current medicines or drug. Recently, they constituted a complicated and elongated procedure which can take many years for completing and generating huge quantities of data.

The procedure itself is put through to specific investigation as its outcomes can either produce or break a current medicine. Therefore, it may also pull apart assertions of producers linked with a drug's/product's abilities and influences on the human body.

The corroboration abilities of the blockchain technique assure that any data saved on a shared log is both obtainable by every possessor and can't be interfered with—neither by the log developers nor by any intermediaries.

4.3   Assertion Judgement

The healthcare industry undergoes ineffectiveness in patient invoicing, with an approximated 5–10% of all healthcare costs being fraudulent because of extra billing and invoicing for assistance which were not issued. Assertion judgement, for example, will be better and effortless on the patient by activating the entire procedure through blockchain. Assertions are normally scribbled in advanced (official) verbalization view, possibly being put through to changing clarification in court. With reference to healthcare, this method is fundamentally inaccurate, because contracts lean to make the situations complex for the purpose of being certainly protectable through a legalized perspective. Rather, they should concentrate to communicate the costs of healthcare assistance to a patient clearly. Smart contracts powered by blockchain have the capability to resolve this issue. Instead of writing in a legal language, the contracts are written in code and they don't leave space for double interpretation. Simultaneously, smart agreements prompt levelness and explorations in the procedure of claim.

4.4   Drug Supply Chain

Healthcare is in continual requirement and can trace its supply chains in their completeness—through a supplier and their producing prerequisites to the end consumer. A producer labels a manufactured medicine—producing a "hash" and transmitting the information on the blockchain. Then, a supplier corroborates the product through the log. Prior to selling it, a dispensary staff can also corroborate the product origin subsequently. At last, a patient buying the medicine can genuinely verify its source and producing background.

Recognizing a fake medicine or recommendation in a short time span not only assists medicine manufacturers/resellers monetarily but may as a matter of fact rescue a survival of patient.

The range of this issue is lurching, as the fake medicine commercial operations have expanded into a $75 billion industry itself. Originalities like MediLedger Project based on the Ethereum, sponsored by industry employers like Genentech and Pfizer, have already developed a drug tracing resolution, concentrating to optimize drug supply chains and removing fake products.

5 Smart Healthcare

Smart healthcare discovers the interactivity among therapeutic professionals and patients as per Du et al [8], medical organizations, and therapeutic gadgets by making health repositories localized therapeutic data outline and utilizing many progressive Internet of Things technique, with the aim that the diagnostic profession steadily obtains information. A significant measure for making therapeutic process smarter and enhancing medical service condition is therapeutic data-sharing. But, the patient data-sharing between institutions is not completely recognized till now, and the great way for resolving this issue currently is blockchain. Blockchain is divided data systems which involve many segregated nodes, which is an advancing technique for distributed and proceedings data revealing between large networks of untrustworthy participants. It characterizes distribution, timecodes, combined supervision, programmability, and tinker-evidence. Blockchain has adequately certain applications in treatment of medical, and the current research generally concentrates to combine blockchain with some information technology for building a unique utilization programme, like utilizing a blockchain technique for building a therapeutic proceeding layer corroboration system; By making use of etheric blockchain of building a therapeutic data revealing policy MedRec which integrates big data with blockchain; and utilizing blockchain technique and the OPAL/Enigma encryption policy for creating a protected surrounding for storing and analyzing medical information. But, the suitability of blockchain technology in the entire smart therapeutic service has inadequate methical experimentation.

In this context, the review will create a customer-focused blockchain intelligent healthcare application system on the basis of contributor hypothesis for exploring its route of establishment. Once all the stakeholders of the system are sorted and analyzed, we obtain the application system on the basis of 10 aspects inclusive of 22 criteria.

5.1   Internal and External Rules

Internal management mostly indicates real-time management of drugs, gadget, and supply chain with the help of therapeutic organizations. External management involves three principles: regulation of medical supply chain, rule of entire therapeutic approach, and rule of therapeutic dissipation treating procedure, and the controlling framework is the rule framework. The application chain is a system which includes services or product transfer through sellers to customers that consists of people, associations, actions, information, and assets. Its aim is to make sure the feature of delicate products during the shipping. Supply chain management is very significant in the service of delivering equipment and products. Also, drug supply chain management is especially significant to track the source of materials used to manufacture, the procedure of producing, and the allocation of accomplished goods. Incorporated supply chain management systems reveal supply chains to fraudulency tinkering. Efficient supply chain management is especially significant in the therapeutic area, in which a menaced supply chain straightly acts on patient protection and health consequences. Blockchain technology is the probable solution for improving protection, integrity, origin of data, and performance of a healthy supply chain. In the supply chain, blockchain technique proceedings are protected and clear and are invariably supervised and registered, enormously decreasing the time needed and also the feasibility of human inaccuracy. Additionally, with the execution of blockchain, the protection of therapeutic items and gadgets can be enhanced. Blockchain technology can be obtained by storing distinctive gadget identifications for every therapeutic gadget and by tracing and issuing firmware upgrades utilizing smart agreements. Therapeutic gadget tracing on the basis of blockchain can also utilize uniformity for preventing misplacement of gadget, stealing, and malevolent tinkering.

5.2   Medical Record Supervision

Supervision of medical record contains electronic therapeutic evidence. Electronic medical record issues an appropriate health evidence saving assistance which can promote virtual computerized access to conventional paper therapeutic evidences, which is an extremely delicate information which has to be revealed between associates, for keeping updated patient background. Recent computerized therapeutic evidences will not assure the protection, secrecy, and accessibility of delicate data, and detection and medicament data of the patients are until now in a condition of distribution in distinct therapeutic organization records. Consequently, patients might lose control of accessible healthcare data; at the same time, medical institutions normally protect primary data management. Utilizing blockchain technology for sharing computerized therapeutic evidence is customer oriented, and patients have the supervision of their computerized, therapeutic evidence. The distributed, protected, and foolproof quality of the blockchain assure the protected repository and transport of therapeutic data and highly decrease the turnabout time to reveal an all-inclusive cost. Also, as patients can take part in their individual health evidences, they will concentrate more on their individual healthcare.

5.3 Therapeutics Enhancement

Enhancement of therapeutics involves two specifications: telemedicine and chosen treatment. The number of therapeutic patients has multiplied theatrically in a lot of nations, building it growingly arduous to patients for obtaining quickest assistance through physicians or nurses. Scientific know-how of telemedicine is viewed as a method for attaining fair and cost-efficient clinical protection. The establishment and familiarization of Internet of Things gadgets and other distant patient supervision systems have given rise to protection threats in the communication and evidence of proceedings of data. On the basis of features of blockchain- safeguarded, unidentified, and clarity, the smart agreement of the block chain is utilized for promoting the analyzation of protection and supervision of therapeutic sensing elements. For instance, in medical chain, an intelligent agreement is introduced; using this, patients can allow doctors to examine medical cases remotely and issue advices or second opinions. Additionally, the data can be securely communicated using blockchain nonsymmetric cryptographic encryption, and the medicine investigation and establishment organizations categorize the data once user authorization is done, for conducting medicine-choosing exploration and establishment.

5.4 Doctor Supervision

The supervision of doctor involves three guidelines: identification, examination of individual, and choice of customer. Blockchain technology permits for enormous interaction between healthcare issuers, patients, and representatives for information interchange, patient chasing, identity guarantee, and verification. Additionally, blockchain can be utilized for recording doctors' treatment and avoiding medical conflicts, hence supporting medical institutions for choosing brilliant doctors and patients for choosing the correct doctors. For illustration, Medico Health is a project based on blockchain which permits totally unidentified and protected patient message along with globe's superior doctors. Data on the validness of doctors' testimonials and permissions is modified in an invariable distributed record, and patient data are only saved and obtained unknowingly by a specified doctor at an agreed time.

5.5 *Therapeutical Assurance*

The blockchain can issue a foremost level of dependability for users because of its Practical Byzantine Fault Tolerance and multi-node co-maintenance nature. France's main coverage group, AXA, has begun providing characterized flight delay coverage on the basis of its Ethereum principles known as fizzyTM, which utilizes smart agreements linked to the globalized air trafficking record for securely initiating reimbursement once flight postponements are found, hence evading the requirement for extra paperwork. A comparable method could be executed for health coverage, closing out the requirement for verifying medical evidences and enormously making the procedure clearer. The use of blockchain technique in medical coverage can decrease the complicacy and price of therapeutic coverage, protecting the attentiveness of patients and reducing the uncollected money of health centers and the price of supervision of coverage organizations.

5.6 *Price Saving*

Price saving incorporates two principles: the price saving of conversation between doctor and patient and the price saving of medicine investigation and establishment. Credit Suisse's review directed in 2016 showed that health centers, medication business, and coverage organizations could rescue funds by executing blockchain technology. One of the primary advantages of blockchain systems and computerized therapeutic evidences is greater approach to therapeutic evidences for both doctors and patients, enhancing nursing effectiveness and standards. Disclosing a log between entities like drugstores and assurance organizations can be exceedingly appropriate for patient costs and supervision of medicines. Especially in the circumstances of supervision of persistent diseases, issuing dispensaries with correct data on medicaments will enhance logistics. Multi-vendor clinical attempts at medicine establishment prerequisites can decrease test costs, while data management systems on the basis of blockchain smart agreement technique can decrease the price to manage multi-vendor clinical tests.

6 Data Protection Storage and Approach Based on Blockchain

The exposure of EMR has caused issues of protection and comfort. Put through to issues of protection, medical data will not be divided explicitly. Certain models based on blockchain are put forward.

Rahul et al. have put forward a "medichain" model which is based on blockchain. This model utilizes the blockchain like a record for storing the entire patient's case information in the block. The evidences of proceedings are hashed for storing acquired hash values in the Merkle tree for ensuring data protection and avert interfering, hence decreasing inaccuracies in clinical making of conclusion. For addressing the issues of a broad span of origins and various layout of medical data, all fields' data are merged into one hyperfield saved in the suggested frame. The procedure utilizes on-chain storage. But, the blockchain is not that much compliant. On-chain storage is costly as well. Wu launches a patient-centered protection-saving entry supervision model into the method of control of accessing of secret information in systems of healthcare. Hence, blockchain technology is utilized for building a secret platform of information depository, and quality cryptographic algorithms are utilized for realizing transmission of information. In this method, the secrecy information is also protected by a file permission agreement for further preventing the stealing of medical secrecy information. The model suggests a triturated secrecy-safeguarding access control procedure which can grant dissimilar rights for members by concluding their categories. Computerized medical evidence information is saved in the cloud database and provided by a mediator cloud servicing corporation. A hash of that data is produced whenever data are saved on cloud. Afterward, hash is saved on the blockchain. Whenever the data present in cloud is tinkered with, they can be contrasted by the value of hash on the chain. The agreement procedure is proof of work (POW) in this model that needs many incorrect calculations through the redistribution points. Liu et al. recommended an insubstantial model on the basis of blockchain to share and protect medical information. The model utilizes substitute re-encoding technique for enabling data dividing between doctors in various health centers. The hash function utilized is difficult to strike. Thus, saved medical information is practically unfeasible to be interfered with. The conventional entrusted verification of notice is enhanced for obtaining a recent agreement procedure which is more protective and dependable. A disease-matchup procedure is created in which patients getting affected from the same disorder can convey with each other. Session keys are set among patients once mutual authentication is done. This procedure can assist patients for exchanging information related to disease. The nonpublic chain is speedy in proceedings but not very distributed. It is more appropriate for applications among organizations or institutions. It is not appropriate in case of a lot of patients and hospitals. Yu et al. launched a EHR sharing plan on the basis of composite chain for storing the nonpublic part of the electronic illustration in the consolidated chain and the public part in the non-private chain. Authorized users will only be able to access the nonpublic part, and the collaborative part can be divided with technological organizations for medical establishment. The model also utilizes off-chain repository, and only data hashes are saved on the chain for preventing data vandalizing, and smart agreements can spontaneously supervise the computerized medical evidence requisition, acceptance, and utilization procedure. The composite chain procedure which the model applies is very unusual. But features are not permitted to the nodes, and ambiguous access control is utilized. Zou et al. have created a structure of a recent chain for avoiding the diverging issues and recommended an agreement approach on the basis of trust for resisting Byzantine attacks. Medical organizations can gather trustworthy points with the help of continual extracting in swapping for computerized medical evidences. The recommended recognition system has to gather recognition scores with a huge number of incorrect computations. A huge quantity of potential is used for obtaining rights of voting. Shahnaz et al. recommended a delicate access system on the basis of blockchain which can grant different rights of permission for patients, doctors, assistants, and supervisors. The permission for electronic illustrations is registered in the model which is put forward, and a full-searchable encryption procedure is utilized for the data on the chain to browse information not having decryption. This procedure secures the secrecy of the data and assures the pace of the inquiry. The procedure also utilizes supervision of access control on the basis of role.

7   The Possibilities of Blockchain in the Healthcare Sector

The healthcare area is an issue-determined, data- and personnel-intense scope in which the capability for accessing, editing, and trusting the data which emerges from its actions are crucial for the activities of the area as an entirety. If the activities are divided among the healthcare area into group, health issue-resolving, clinical resolutionproducing, knowledge and evaluation of awareness-based supervision; obtaining the required health results depends upon taking a comprehensive group of health personnel which employ the highly suitable awareness, techniques and abilities while handling the patient. When cooperating with instructional organizations, the healthcare area needs to issue access to patients and issue a domain to train so that students can succeed and improve the required abilities. In exchange, the instructional organizations issue the area with knowledgeable staff. While cooperating with organizations with an investigation and engineering schedule, health organizations need to help in issuing access to experts, authorities, test persons, and illustrations. While engaging in intended clinical tests, health institutions have to help to develop, plan, conduct, and report the investigation. In exchange, the investigation and engineering organizations issue the healthcare area with revised awareness, procedures, and mechanisms. Thus, the tasks of health associations are firmly interfaced with associations occupied in instructing health personnel and in biomedical investigation and engineering. The actions need effectual exchange of agreements, data, and proofs related to patient, and methods of compensations, which successively means interchanging data over organized boundaries. Simultaneously, health institutions are chosen for protecting the most confidential data that patients select to divide with them.

For maintaining the secrecy of patient and interchange of information with remaining organizations in the healthcare environment, access control, source, integrity of data, and coordination are critical. The conventional manner to achieve access control generally accepts faith among data's owner and the systems saving them. These systems are frequently servers completely committed to define and enforce access control procedures. Coordination is the capability of various information systems, gadgets, or applications for connecting, in a proportionate way, between and over organizational restrictions for accessing, exchanging, and collaboratively utilizing data between stakeholders, with the aim to optimize the health of persons and communities. Data origin pertains to the archival evidence of data and their sources. In the data of health area, sources will be able to, for instance, deliver verification and clarity in computerized health evidences and for achieving faith in computerized health evidence software system. Data integrity as a common interpretation which Courtney and Ware gave is the data standard interpretation that handles the predicted data standard. This means that the level to which the predicted standard of the data is met or exceeded finds out the integrity of data.

Healthcare institutions recently encounter a growing requisite of actual data through industry and investigation corporations. Simultaneously, unauthorized sharing and extremely published theft and burglary of confidential data constantly spoil the general faith in healthcare organizations. A third issue is carelessness among the healthcare surroundings that makes use of the very same faith (like the issues with fake drugs, methods, capabilities, and patients). Overall, this is a scenario which directs reconsidering and thoughtfulness of another method. Using certain essential characteristics like decentralization, distribution, and data integrity, and in the absence of any required third party, blockchain technology has a lot of attractive characteristics that can be used for enhancing and obtaining a higher degree of interactivity, sharing of data, access control, origin, and data truthfulness between the recommended collaborators, besides going toward a current framework to build and maintain faith.

8 Blockchain Use Cases in Healthcare

8.1 *Blockchain for Enhancing Medical Evidence Access*

As per Weinberg et al [10], the predominant accepted healthcare instances for blockchain are management of patient data. Health agencies have the tendency to segregate the medical evidence, making it impractical for finding a medical history of patient by not asking their earlier supervision issuer. This method can take an important part of time and may frequently cause inaccuracies because of human mistakes.

Evolved on the blockchain of Ethereum, MedRec is a "system which focuses on patient action, issuing a clear and approachable perspective of therapeutic background." MedRec is considered for storing every patient's data in a location, making it easier for patients and doctors for viewing. In its recent design, issuers sustain the blockchain using the mechanism of proof of authority (PoA).

8.2   Blockchain in Healthcare: Cutting Costs

SimplyVital Health has two assignments functioning on the blockchain technology. ConnectingCare, as stated by CTO of SimplyVital Health Lucas Hendren, "utilizes safe-keeping collaboration and financial predictions to assist issuers in packed-up remunerations get awareness into patients' conditions when they check out from the health centers." It is presently on the market, assisting healthcare issuers to find how much a patient's supervision will fetch them when packed up with so many organizations.

SimplyVital Health could gather data on the blockchain and find the requirements for their subsequent assignment, Health Nexus after the release of ConnectingCare. Health Nexus saves information of a patient on a blockchain for every party for viewing. It also will feasibly permit patients for selling their data to investigators for benefit.

8.3   Enhancing Medical Evidence Retaining Using Blockchain

The Taipei Medical University Hospital and Digital Treasury Corporation (DTCO) have newly revealed Phros. It intends to grow clarity among therapeutic organizations by placing every patient's therapeutic data on a blockchain.

It consists of pictures and a lot of information which concerns a patient's state. The information can be obtained by doctors and the patients themselves using a mobile app. It also enlarges the protection of medical information all over the decentralized ledger technology (DLT).

8.4   Blockchain and Healthcare Can Be Utilized for Actions of Release

Utilizing Hurricane Maria release actions as a position of source, the Department of Defense (DD) in concurrence with the Defense Logistics Agency (DLA) investigated the capability of blockchain for subsequent release actions. At present, the DLA follows logistics using centrally supervised systems that can be divided among numerous organizations. This makes it tough for release actions to cooperate effectively. A system based on blockchain would permit data to be appended and followed using a log, besides issuing a live feed of release actions of many sectors. It has the capability for saving lives and money.

8.5   Blockchain for Stopping Fake Drugs

It is approximated that tens of thousands of people lose their lives every year because of fake drugs. FarmaTrust plans to terminate that using blockchain.

Blockchain of FarmaTrust will be divided into four various parts. Regulatory compliance assists pharmacy corporations by assuring they are functioning in suggestions which the government has established. Track and trace utilizes the blockchain to managing inventory each time it goes. Supply chain visibility follows when a drug is varied in any manner. In the end, consumer confidence app permits customers for seeing the life process of their medicine.

8.6   Make Computerized Health Evidences Easily Obtainable Using Blockchain

Computerized health evidences can be difficult to control. One healthcare issuer's computerized health evidence for a patient can vary from the same patient's other issuer. MTBC proposes to vary that by utilizing application program interfaces (API) and blockchain. The objective is to put supervision in the patient's hands. A patient can have the capability to permit the records' transmission among doctors. The blockchain API executes on the hyperledger principle and is recently obtainable.

8.7   Following Clinical Experiments and Medicines Using Blockchain

Currently available on their website is a range of blockchain platforms from multinational supply chain manufacturer for the Internet of Things, Ambrosus. The business is keen to increase their applications for the pharmaceutical sector, nevertheless. The organization is recently in discussions with enormous medicine providers regarding introducing a pilot program using their new products.

9   Problems of Blockchain in Healthcare

It will be hard for applying a current technology like blockchain in healthcare on an important scale. A few problems blockchain could encounter are as follows:

- Dissimilar topographical places have various sets of medical recommendations.
- Unpredictability in finding who needs access.
- The highest amount of data can be saved in a single block.

- If blockchain technology is utilized in the healthcare industry, every medical evidences, documents, photographs, and lab reports will be saved on it. The merged size of all of these files might effortlessly overreach the storage capability of recent blockchain technology, producing a difficulty.
- There is recently no procedure in place to find who possesses the healthcare data and who approves authorization for sharing it.
- There is no understandable approximate of how much it will cost for setting up and maintaining blockchain technology in healthcare.

10   Transforming and Enhancing Healthcare Industry, as per Chirag et al [12] Using Blockchain

There are a lot of advantages which blockchain technology fetches to the healthcare industry, involving protection of data, supervision of data, etc.

10.1   Enhanced Protection of Data

Blockchain is a distributed database which permits for protected, clear, and safeguarded healthcare data management. Blockchain technology assists to resolve most of the usual problems when it is utilized in healthcare, from information leakage to interactivity concerns deceit.

By implementing blockchain in healthcare, patient data is saved in a distributed network instead of a central server. This becomes harder for hackers for accessing or tampering medical data. Additionally, blockchain permits various healthcare issuers for sharing data with each other while sustaining secrecy of patients.

Blockchain data is encrypted and can be retrieved only with approved authorizations. Eventually, blockchain assists to decrease corruption and fraud in the healthcare industry by issuing an inspected sequence of every transaction.

10.2   24/7 Supervision of Data

Continual supervision of data is an essential specification in the healthcare area regarding administration and enhancement of electronic health records. Many human existences are at risk daily; a minute inaccuracy gap can have catastrophic results.

Beginning from the readings of blood pressure of a patient to the prescriptions of doctor, each data bit associated to a patient's health has to be supervised and retrieved 24/7. But the enduring healthcare support is not able to issue this extent of obtainability of data because of several causes like interaction problems, unintentional information disclosure, etc.

The blockchain technology can issue 24/7 obtainability of data and assist to enhance electronic health record management with its allocated character. Additionally, blockchain assists to resolve interaction problems by issuing a systematized principle for data-sharing among various healthcare issuers.

10.3   Enhanced Clinical Experiments

According to the review, the worldwide clinical experiments market area was estimated at USD 38.7 billion in 2021 and is predicted to expand to 52.0 billion by 2026, with a CAGR of 6.1%. Numerous characteristics such as the rising number of clinical experiments, increasing insistence to outsource R&D, and drug promotion are handling development of this market.

But the clinical examination method is repeatedly long and complex, and it involves a lot of stakeholders. This can cause incompetence and postponement that can influence the affluence of clinical examinations. Additionally, clinical examination data is often stored, making it harder for sharing information.

Blockchain technology has the capability to assist in resolving these issues by issuing protected and clear way for storing and managing clinical examination data. Every measure of the clinical examination method would be registered on the blockchain, making it effortless for following data and identifying inaccuracies. Additionally, blockchain can be utilized for corroborating the recognition of patients and doctors, assuring that only permitted individuals have access to clinical examination data.

10.4   Enhanced Protection of Drug

Unfavorable drug reactions are a crucial issue in the healthcare industry. They are in control for thousands of treatments and deaths every year. Additionally, the cost to treat unfavorable drug reactions is approximated to be billions of dollars per annum.

A blockchain technology database can be utilized for tracking the origin of drugs all over the supply chain. This can permit for the quick diagnosis of fake or contaminated drugs. Blockchain can be used for tracking results for patients, permitting for the quick identification of unfavorable reactions. Enhanced drug protection is one way that blockchain can potentially modify the healthcare industry.

10.5   Smart Agreements

Smart agreements in blockchain can possibly integrate various nonidentical processes in the healthcare industry. For instance, they can be utilized for spontaneously processing insurance assertions, corroborating acceptability for advantages, and planning commitments.

Furthermore, smart agreements could be utilized for managing patient data and evidences, assuring that only permitted individuals have access to sensitive information.

10.6   Claims and Billing Management

Management of assertions and billing is the role of the process where ordering and clarification of the medical assertions associated with the patient's detection, medicines, and therapies are concluded. There have been various events in which the medical records were sacrificed, causing cases of deception and robbery.

Blockchain makes it convenient to eradicate these illustrations because the technology performs all over documenting medical data and saving it in the open digital ledger in a way that when any variation is done in the information, it is clear to everybody who is a portion of the blockchain.

There are various other benefits of utilizing blockchain, such as decreasing medical inaccuracies, improving public health, etc.

11   Reasons for Using Blockchain as per Anon et al [14] for Healthcare Sector

Blockchain is a distributed log system which can record transactions in a database made up of blocks in which blocks are chained together with the hash value for forming a chain-like structure. Blockchain has a lot of characteristics which make them a good selection for the healthcare systems:

- *Protected:* Blockchain is a protected network of structures. When a block is adjoined to the blockchain, it will need the concurrence of the most of remaining nodes in the network for altering the data in the block.
- *Decentralized:* Blockchain is a decentralized network of nodes in which no one needs to believe or realize others in the network. Every member of network has data replica like remaining nodes in the network. In case a member's log is modified, then most of the member nodes of network will reject it.
- *Self-rectification:* If data present in any one of the blocks in the network is varied, then other nodes in the network verify their blocks, and note each other to remove the inaccuracy.

- *Distributed*: Blockchain is a distributed ledger in which self-reliant computers document, divide, and coordinate the proceedings in their corresponding electronic logs.

11.1   Advantages of Blockchain in the Healthcare Area

Protecting data of patient: Protecting patient data is the most crucial part in the healthcare area. Damaging the patient records can cause issues for identifying the disease or issue in the patient through physicians and health centers. From 2009 to 2017, about 176+ million patient evidences were infringed. The hackers peculated the bank and credit card information and utilized them in corrupt ways. Blockchain helps to access data effortlessly and in a more protected manner as it is distributed, tamperproof, and trustworthy.

Supply chain management of medical drugs: Drugs or medicines are not produced in health centers. They are produced in pharmaceutical organizations and laboratories all over the world. Those drugs are further allocated over the countries as per their requirements. The therapeutic supply chain needs to be foolproof and clear for the wholesalers and shippers. Blockchain eradicates this issue as it has characteristics like clarity, distribution, and being safeguarded. When a decentralized log is generated for the drug, every transferring position will be entered into the blockchain through source position to terminus making it clear over the shipment.

Single longitudinal patient records: All patient record types are entered into blockchain log. Evidences like archives of ailments, outcomes of laboratory tests, payment for therapies or treatments, etc. assist to predict the ailment on the basis of patient's consequent visit with the help of machine learning and artificial intelligence. Utilizing the pre-collected evidences assists hospitals to offer cut prices to their customers by inspecting their evidences. It will also assist to master patient indications making records well-organized scrupulously and avoiding inaccuracies.

Supply chain optimization: The main ultimatum in the healthcare area is to assure the reliability of the source of medicinal products for ensuring the genuineness of drugs. Utilizing blockchain technology, the foods can be detected through producing to all phases of the supply chain. This permits the thorough clarity of the products for buying.

This assists organizations to apply AI and forecast stipulation better and enhance the supply as a consequence, and simultaneously, customers' trusts can improve.

Trackability of drugs: Blockchain is the most authentic, reliable, protected solution for tracking all drug to its sources. All blocks of data consisting of drug information will have a hash value associated with another block of data. This hash value assures that the data is not tinkered with. The blockchain proceedings are conspicuous to all permitted parties. Drug purchasers can assure the genuineness of the obtained products by examining the QR code and scrutinizing all the required information like producer's specifics.

Cryptocurrency remittances: Blockchain makes it feasible for receiving medical help and remunerates for those utilities in cryptocurrency. For instance, Aveon Health and Micropayment are the technology which is blockchain-based and utilizes the benefits of blockchain technology. Aveon Health utilizes the benefits of Bitcoin cryptocurrency. The micropayment model will register all factors of the patient's actions associated with the treatment investigation.

Distributed storage of medical records: The interplanetary file system (IPFS) permits saving interpreted file data in distributed repository surroundings. The file system creates a single hash known as base configured IED (intelligent electronic device) representation that assures that all the files' interpretations are similar. IPFS returns hash and hash permits users for accessing data on websites. Blockchain files are persistent and websites vary frequently and DNS needs to be varied after each update. Hence, IPFS resolves this issue when site addresses are mapped to DNS.

Update medical supply chain management: Blockchain's protection, dependability, and distributed storage make it suitable to manage and monitor the operation of drug products. It assists to build a trustworthy network of sellers for improving patients' protection. Blockchain incorporates all the procedures such as production, packing, allocation, shipment, and repository information in a single fixed record which is saved protectively.

Enhanced computerized health evidence systems: Computerized health evidence systems are the health data in computerized format which are produced and saved by many healthcare facilities. Blockchain details drawbacks such as obtainability, interactivity, and authorization by relating computerized health evidences and dividing possession of the evidences between all associates.

Enhanced recruitment for clinical examinations: Investigators have produced a blockchain on the basis of Ethereum; the blockchain counterfeits the procedures for recruitment. The consequential blockchain-based system permits every researcher for viewing analysis information while securing the secrecy of the practice members.

11.2 Use Cases for Blockchain in as per Arunda et al [15] Healthcare

Blockchain technology has a broad scope of potential scenarios in the healthcare industry. These scenarios include the following:

- *Electronic health records*: Blockchain technology can be used to protectively save and share patient health evidences. This can help to reduce the risk of data breaches and assures that all data is correct and updated.
- *Clinical trials*: The healthcare providers leverage the blockchain to preserve the clinical trial data and share data with clinical trials more securely. Thus, this scheme reduces the cost associated with the clinical trials and saves the time.
- *Research*: The research studies state that the storage of the data and exchanges of the data are more secure through blockchain technology. Further, ensure that all the data are up-to-date and accurate.

- *Insurance claims*: The insurance claim-related data are preserved, and exchanges among the insurer and agents are more secure through blockchain technology resulting in time- and cost-savings.
- *Prescription management*: The prescription data stored and exchange among healthcare practitioners are secured by leveraging blockchain technology resulting in risk mitigation and eliminating fraud.
- *Medical device data*: The medical devices extract the patients' data which is securely stored and shared with medical practitioners through blockchain technology. This can help to ensure that all data is accurate and up-to-date.
- *Supply chain management*: Blockchain technology is used to protectively store and share data related to the supply chain. This can help to reduce the risk of counterfeit drugs and ensure that all drugs are up-to-date.

11.3 Use Cases That Can Transform Healthcare Sector as per Gondhale et al [13]

- Supervision of patient health evidences

Earlier health evidences of patients, their medicines and treatment, treating specialist, etc. are commonly saved as printouts, and the patient needs to carry those papers to their every hospital review.

Resolution. It would be convenient to access the information anywhere and anytime if the health evidence data is uploaded into the blockchain. It ensures the dependability on the source. As the blocks of a blockchain are encrypted, the data cannot be altered.

- Medical Staff Credential Corroboration

It is troublesome with patient records and also with medical staff credentials. Corroboration of certificates, license, address, and id proofs for ensuring accordance and genuineness is a time- and energy-captivating procedure.

Resolution. This corroboration procedure becomes easy when all the documents are digitized and saved in blockchain. The personnel whose information must be corroborated can make use of their fingerprint or retinal scan for accessing their documents and hence assuring protection.

- Fake Drug Issue

Fake drugs are in the market since long. They not only cause danger to the consumer's health but also put the authentic brand's repute under hazard. The fake drug market is approximated to be a USD 200 billion industry.

Resolution. Blockchain's track and trace solution tracks drugs' information starting from its production to consumption. Uniqueness or duplicates can be easily traced. Hence, the fake pharmaceutical products can be checked using trackability of blockchain.

- Transactions and Cost Cutting

The continual process in the healthcare area is transaction. Medical products, equipment, patient and staff transactions, etc. are done on everyday motives. These proceedings are done with the cooperation of trustworthy mediators which charge an additional proceeding cost and in certain times delay proceedings.

Resolution. Transactions based upon blockchain remove the requirement for such mediators, hence excluding proceeding cost and time. The proceedings are fast and can be completed in few seconds or minutes over numerous principles.

- Patient id and guidelines for corroboration for Insurance

Medical treatments can be costly and so patients opt for health coverage. However, demanding insurance could incorporate corroborating patient details, specifications and status, hospital and insurance strategies, and the suitability of the patient relying on the kind of treatment.

Resolution. These corroborations can be done effortlessly, and it saves energy and time, using blockchain intelligent agreements. Intelligent agreements are programmable digital contracts among two or more parties which implement the desired task, once all approved features are satisfied. Thus, on the basis of insurance terms and conditions and patient evidences, insurance claims can be perfectly corroborated.

- Supply Chain Clarity

Clarity in the supply chain of healthcare industry goes a very long way to gain customer's faith and make repute. Following every record of some product right from its producing phase to the level of depletion assures this clarity.

Resolution. Supply chain traceability is the well-known use cases in healthcare. Blockchain traceability comes into the picture, with that being said. Blockchain's track and trace solution ensures supply chain clarity.

- Enlarged Compliancy

The healthcare area is put through a lot of rules and regulations through the governing bodies. When the ultimatum regarding supervision of records, documents, and other things is given, assuring compliancy becomes tiresome. There is always a feasibility of inaccuracy because of carelessness.

Resolution. All proceedings and evidences are saved in real time and are easily obtainable with the help of blockchain, hence growing clarity and compliancy.

- Supervising remotely utilizing IoT Sensing elements

Supervising patient vitals from time to time and safekeeping evidence of the same need a team of practitioners to be constantly on field. Utilizing personnel for certain things as little as this is not worth the time and energy.

Resolution. Using IoT sensing elements, the vitals can be effortlessly supervised and evidenced, and the details can be saved in blockchain together with other patient evidences. This permits the personnel to be assigned to more significant projects.

- Networking and Association

Associating with medication, laboratories, and other resource issuers is a united part of the healthcare industry. Associations are feasible only with faith among the groups. Thus, agreements are problems.

Resolution. Groups can model a customizable agreement which gets implemented when agreed upon terms are satisfied using blockchain's smart contracts. This sets up faith among groups and assures equitable business.

- Management of Resources

Resources, in terms of personnel and skillsets, need to be supervised for smooth functioning.

Resolution. Resource supervision becomes easy using records based upon blockchain.

Solutions based on MSRvantage's blockchain is advantageous for the healthcare industry broadly. Blockchain assures the removal of third parties in proceedings, decreases medical inaccuracies, and builds faith with the consumers. The future of healthcare sector is blockchain.

12 Problems Facing the Healthcare Industry

12.1 Interaction of Data

Therapeutic data interaction delineates the capability of personnel or systems for sharing and exchanging therapeutic evidences. The interchange is critical for data integration. Healthcare issuers can decrease unchanging duties, save time, and enhance standard of healthcare by getting access to the evidences.

However, there is a usual problem to attain accurate interaction among healthcare systems and issuers. The utmost hindrance is occlusion of data, because certain sellers force charges for data transferring out of their system. Another significant hindrance is the absence of quality and structure to share and manage the data over health systems. The incapability for measuring or analyzing this data leads to postponement and incorrectness. Accurate interactivity can only be feasible if lucidity of data is obtained and supported.

12.2 Drug Thieving or Faking

Thieving or supplying fake drugs produces an enormous fear to the healthcare industry. Investigation by the World Health Organization approximates that over 10% of edibles and medicines around the globe are fake. Fake drugs suffer monetary impoverishment to producers. It can also obstruct further investigation and innovation for enhancement or, worse, cause death to the patient.

Discovering fake drugs is tough as many products pass through a prolonged marketing route. This establishes supply chain entrance.

12.3 *Data Evidence Fraudulence*

Data evidence fraudulence occurs anywhere and anytime. The goal of the most of healthcare issuers is to forbid this by having data storage replica. But what happens when the drive malfunctions?

A current overview proves that many data leakages in the healthcare industry arise through hacking or IT-related occurrences. Fraudulence can be caused by unapproved approach.

Medical evidence necessitates the patient's essential information—name, date of birth, employment history, etc. It is unfavorable if it falls into incorrect hands. Burglars will utilize the data for creating incorrect identities, documents, or prescriptions.

12.4 *Integrity of Supply Chain*

The COVID-19 pandemic endangered the inadequate flexibility of healthcare supply chains. This arises due to insufficient access to centralized data.

The inadequacy and unpredictable requirement for medical stocks cause logistical issues. It frequently comes from an inadequate centralized and real-time data through siloed systems. As it is unfeasible for tracking numerous data, approximating the extent of the requirement for medical goods is inconvenient.

Blockchain is not a remedy for data systematization as per Deloitte. However, it issues a favorable recent decentralized organization which reinforces the consolidation of healthcare information between users.

13 How Blockchain Issues Solutions to Healthcare Problems

Blockchain has been the crucial go-to in instances of genuineness and clarity. Assume an instance in which stakeholders can obtain information when they need it. Unquestionably, this system can be the solution to coordination in the health industry.

Accomplishing accurate interactivity begins with unopposed and restricted data access. Using blockchain cryptographic private/public key access, decentralized data, and proof of work/stake, healthcare evidences can be obtained with no theft fear.

Using this, doctors can produce instructed and faster conclusions in managing protection. Remaining healthcare professionals can obtain patients' information, striking out locality and third-party hurdles.

Products can be detected and corroborated using blockchain. It becomes convenient to trace and correct problems. Sequentially, it enhances the protection of the therapeutic supply chain network.

For example, a pharmaceutical company can trace all the details of a drug, from production to supply. This is feasible when transaction is saved on the blockchain.

Hence, is it simple to determine deceit and where it arose on the chain. This information assures the acceptability of the product to the pharmacist and the patient.

Health institution can activate medical supervision using intelligent agreements. This assists to make the processes fast and enhance effectiveness. It also removes the probability of human-based inaccuracies.

13.1 Blockchain Technology Will Issue Impulsion to Healthcare 5.0 in 2023. How?

Blockchain is an apportioned flexible log to record transactions, trace properties, and build faith.

An "apportioned…log" specifies a computerized record which is observable to every group which comes up to appending recordings to it. Whenever a new "block" of information is appended to the "chain," the blockchain gets updated. The modified blockchain is spontaneously shared to all groups. The groups can be thought of as partners and cofounders of this distributed data. Assume computerized therapeutic evidences for a patient.

A computerized therapeutic evidence present in blockchain is modified by physicians who treat the patient, and patient also gets access to modify. Sequentially, the patient will be able to modify any third-party diagnostic examination outcomes for the doctor for reviewing.

An "…apportioned log" specifies the characteristic that a block in a blockchain cannot be updated once it is formed. The data present in blockchain consists of distinctive computerized identifications which are formed by a method known as "hashing."

Any effort for changing data can be pointed out fast. Furthermore, the decentralized character of the blockchain makes it difficult to replicate any unapproved variation on every distributed category. In example of computerized therapeutic evidence, this unchangeable feature builds computerized therapeutic evidence on a blockchain dependable for health center controllers allotting doctors and for medical insurance organizations who require clear remittance information.

A computerized procedure for "registering proceedings" in divided and unchangeable ways is advantageous to a broad scope of contributors in healthcare conveyance. Instances are given below:

1. Using blockchain, a medical school will be able to issue degrees. A graduating student and also her engaging health centers can obtain degrees instantaneously, hence creating spontaneously corroborated testimonials.
2. An enterprise who builds a clinic can corroborate possession of land through land history blockchain. Such a blockchain consists of records of all preceding land holders and records any concerns in selling or movement.
3. A researcher who conducts a clinical examination can query engaged health centers for forming a blockchain, hence facilitating collected data integrity and also forming a divided storehouse which can be scrutinized by investigators.

While referring to "tracing properties," blockchain can provide on a small timescale. Here, the plan is that a blockchain is built up when an item moves from one place to another, every position appending its data.

Presume a medicine supply chain. This starts from a production point through which the product is delivered with the help of feasibly many storehouses utilizing many transportation vehicles. Ultimately, it arrives a drugstore, either through supply or within a clinical prerequisite. From there, patient can obtain it as a medicine. Tracing a product between this transfer is beneficial for management of catalog and expensive for organization of treatment like any supply chain.

Feasibly most significant of everything is "establishing faith." Blockchains are created for functioning in clear and protected manner. This suits effectively into the healthcare needs, in which considerable understanding of faith is needed and is in fact presumed frequently. In the situation of pharmaceutical supply chain, it is presumed that the production point has the accurate registrations, for example, with the US Food and Drug Administration (FDA), and that the medicines it produces have the correct agreements, for example, through the Drug Controller General of India (DCGI) organization. A faith is placed that the medicine is not substituted by a fake drug either during transport or at the drugstore.

A blockchain assists to corroborate everything and maintain faith in the procedure. Faithful data and protected digitalization will be solutions because healthcare 5.0 is set to enlarge. However, there are issues to overpower. The most significant of these comprises extent. The procedure of hashing is measurably intense and exhausts efficiency. More effectual calculation will become essential for blockchain to increase over crypto. The other conditions of enlargement are the dimensions and character of healthcare data. Agreement needs from HIPAA and GDPR connect to the complications to store huge data quantities protectively on a blockchain.

In the years ahead, other decentralized logs can appear as substitutes to the blockchain. For example, the organization Hedera utilizes a substitute known as hash graph for writing smart agreements. SAFE is a computerized health principle through Mayo Clinic which has utilized the technique to corroborate COVID-19 test and health condition. Additional computerized health implementations are developing on the basis of smart agreements among issuers and recipients of healthcare. Blockchain and its successors are establishing for making healthcare superior to everybody.

14 Pros, of Blockchain in Healthcare

14.1 The Key Differences Between Blockchain and Conventional Data Management Process

In a conventional record, the data is kept on one intermediate server (or network of servers) with an incorporated record manager. Instead, blockchain is a procedure to manage data where there is an electronic log attached that is decentralized over a peer-to-peer network in the absence of central management of the data.

14.2 Blockchain Adjusts to Scenarios in Which the Clarity and Invariability of Data Are Required

Blockchain is most appropriate to record proceedings with a lightweight fingerprint, where clarity and invariability are the benefits. Also, it is better appropriate to scenarios in which there is less faith between the members of a network and the measurement of the data blocks is adequately low. Scenarios in healthcare in which blockchain can be significantly beneficial comprise corroborating the individuality of patients or vendors, supply chain management, and supervision of the active allowance of the patient for data utilization.

14.3 Supervision of Patient Authorization and Data Access Allowances

Blockchain can be an auditable and clear practice for people, utilizing their distinctive identifications and encryption key, for permitting other parties for accessing your individual health data. This incorporates permitting authorization to healthcare professionals, service issuers, and other applicable actors (e.g., investigators and societal protection issuers) for accessing your medical evidences and other information for the aim of issuing continuous healthcare or to allow exploration, performance, or other non-primary utilizations of your data.

Since electronic data can be utilized and repurposed perpetually, and recent exploration matters and objectives for data utilization arise interminably, blockchain-empowered progressive or "active" approval issues a very purposeful alternate to "typical or one-time" approved models. If a person determines to vary their permission or consent terms, the variations can be included as a new block which overrules the earlier recommendations in the chain.

15   Drawbacks of Blockchain in Healthcare

15.1   *Establishment Cost of Blockchain*

Though the cost to develop, sustain, and improve a healthcare blockchain is not known, there are many US web establishment companies or US convention software establishment organizations which impose by themselves. But it is unknown about the market quality.

The affective character which influences the general conclusion to execute blockchain technique in the industry of healthcare is cost.

15.2   *Blockchain Is Not Able to Go Behind: Data Is Immovable*

One of consequential drawbacks of blockchain is data immovability. It is fair that many systems gain from it, incorporating the supply chain, monetary systems, etc. But, if you view the functioning of networks, you have to analyze that this immovability can only be available when nodes present in the network are clearly decentralized.

An entity can control a blockchain if it possesses more than 50% of nodes and becomes endangered.

The data cannot be deleted once it is written. Every individual should have secrecy rights. But, if the same person utilizes a digital platform running on blockchain technology, he cannot remove his trace from the system if he does not want it there. Simply put, there is no process that you can detach the trace of it, leaving the secrecy privileges in parts.

15.3   *Proficient Awareness*

Executing and supervising a blockchain project is hard and thus needs extensive awareness of the organization with a view to going through the whole procedure.

It is required to hire many proficient in the area of blockchain, which is a problem and thus counts as one of the drawbacks of blockchain. Also, the current proficient should be trained to use blockchain and make sure that the management team can follow the complications and results of a business powered by blockchain.

They can analyze your needs and assist reconstruct your commercial operations procedures to use Blockchain in this manner.

In addition, blockchain establishers and experts are more difficult to get and hence will cost more in comparison with conventional establishers because of their supply and demand ratio.

16 Conclusion

The function of blockchain is to register every kind of transactions in a distributed log, apart from remaining structure of management. It is correct and effortless; saves cost, time, and effort productively; and consequently liberates effort of management. The considerable issue which the healthcare industry overlooks is the revealing of important data and utilization for malevolent gadgets and other particular interests that this technology applications can sort out fast. The other area of significance is permitting users and parties to access the database to the recent updated and genuine records of patients and considerations. The scope of blockchain of healthcare looks super up-and-coming and exciting as it contributes to solve certain industry's pressing problems.

References

1. Hasselgren, A., Kralevska, K., Gligoroski, D., Pedersen, S. A., & Faxvaag, A. (2020). Blockchain in healthcare and health sciences—A scoping review. *International Journal of Medical Informatics, 134*, 104040.
2. Alla, S., Soltanisehat, L., Tatar, U., & Keskin, O. (2018). Blockchain technology in electronic healthcare systems. In K. Barker, D. Berry, & C. Rainwater (Eds.), *Proceedings of the 2018 IISE annual conference*.
3. Prokofieva, M., & Miah, S. J. (2019). Blockchain in healthcare. *Australasian Journal of Information Systems, 23*. Research Note Blockchain in Healthcare.
4. Haleem, A., Javaid, M., Singh, R. P., Suman, R., & Rab, S. (2021). Blockchain technology applications in healthcare: An overview. *International Journal of Intelligent Networks, 2*, 130–139.
5. Zaabar, B., Cheikhrouhou, O., Jamil, F., Ammi, M., & Abid, M. (2021). HealthBlock: A secure blockchain-based healthcare data management system. *Computer Networks, 200*, 108500.
6. Xi, P., Zhang, X., Wang, L., Liu, W., & Peng, S. (2022). A review of blockchain-based secure sharing of healthcare data. *Applied Sciences, 12*(15), 7912.
7. Tandon, A., Dhir, A., Islam, A. N., & Mäntymäki, M. (2020). Blockchain in healthcare: A systematic literature review, synthesizing framework and future research agenda. *Computers in Industry, 122*, 103290.
8. Du, X., Chen, B., Ma, M., & Zhang, Y. (2021). Research on the application of blockchain in smart healthcare: Constructing a hierarchical framework. *Journal of Healthcare Engineering, 2021*.
9. Razzaq, A., Mohsan, S. A. H., Ghayyur, S. A. K., Al-Kahtani, N., Alkahtani, H. K., & Mostafa, S. M. (2022, December). *Blockchain in healthcare: a decentralized platform for digital health passport of COVID-19 based on vaccination and immunity certificates.*

10. Weinberg, B., & Weinberg, B. (2020, August 21). *14 major real use cases of blockchain in healthcare| OpenLedger Insights. OpenLedger Insights.* https://openledger.info/insights/blockchain-healthcare-use-cases/

11. Kanika K (2021, December), *What is Blockchain in Healthcare and its Top 6 Applications* https://www.terasoltechnologies.com/blog/what-is-blockchain-in-healthcare

12. Chirag. (2024, April 23). Breaking Barriers: Blockchain's Role in Reshaping Healthcare. *Appinventiv.*

13. Gondhale, S. (2022, March 29), *10 Blockchain Use Cases that Could Transform the Healthcare Sector!*

14. Sharma, T. K. (2019, January 16). *Importance of blockchain in healthcare data management | Blockchain Council.* Blockchain Council. https://www.blockchain-council.org/blockchain/importance-of-blockchain-in-healthcare-data-management/

15. Arunda, B. (2022, December 21). *The Potential of blockchain in healthcare: Explore use cases.*

Innovative Solutions for Sustainable Medical Services: A Look into Quantum and Blockchain Technologies

Reena Thakur, Parul Bhanarkar, Uma Patel Thakur, and Mustapha Hedabou

Abstract Blockchain has now become a center for digital transformation. Due to its many industrial benefits and potential, it becomes the core of developing primary sustainable services. Adding up more security to the data has facilitated more informed decisions when using the application in the medical field. Research has shown that an active and preventive healthcare approach requires meaningful insights and massive data for better decision-making. The integration of quantum and blockchain technologies will bring about advancements in healthcare systems, fostering their evolution. The work mentions the challenges to be addressed while developing better intelligent healthcare systems. The work has introduced recent developments in quantum and blockchain technologies in healthcare management systems.

Hopefully, this research will serve as an appreciated resource for a widespread range of interested parties, raising their empathy of the possible benefits of employing quantum and blockchain technology for sustainable medical services and encouraging further systematic investigation into this area. Additionally, the significance of blockchain is examined, and it is found that healthcare systems will use blockchain technology and the associated processes. By enabling the secure and decentralized storage of substantial data and ensuring access whenever and wherever needed, this technology greatly streamlines operations. Quantum blockchain enables the advantages of quantum computing, including the speed and ability to gain thermal imaging established on quantum computing. Another mechanism for preserving the privacy, reliability, and data availability records is quantum blockchain. This chapter also explains how blockchain might affect sustainable perfor-

R. Thakur (✉) · P. Bhanarkar · U. P. Thakur
Department of Computer Science and Engineering, Jhulelal Institute of Technology, Nagpur, India
e-mail: en19cs601002@medicaps.ac.in

M. Hedabou
School of Computer Science, Mohammed VI Polytechnic University, Ben Guerir, Morocco
e-mail: mustapha.HEDABOU@um6p.ma

© The Author(s), under exclusive license to Springer Nature Switzerland AG 2024
S. Pulipeti et al. (eds.), *Quantum and Blockchain-based Next Generation Sustainable Computing*, Contributions to Environmental Sciences & Innovative Business Technology, https://doi.org/10.1007/978-3-031-58068-0_8

mance by enabling data security, traceability, real-time data sharing, and transparency. If blockchain and quantum computing are combined, it may be possible to handle medical records more quickly and privately. The main topic of this chapter is quantum technologies. Examples and benefits of quantum computing, AI, and nanotechnology are provided, along with applications in the medical field. This chapter was developed to serve as a technical resource for people working in various practical contexts and application areas, including those in the academic and professional communities and those in positions of decision-making authority.

Keywords Healthcare · Quantum computing · Blockchain technologies · Blockchain applications

1 Introduction

Blockchain, an open-source and decentralized digital ledger, serves as a record-keeping system for transactions across a multitude of computers. In this system, altering past records necessitates the modification of subsequent blocks as well. A blockchain is essentially a vast network of interconnected and verifiable "blocks," which is why it's aptly named. What makes blockchain particularly valuable is its unwavering accountability; each transaction is meticulously recorded and accessible for public scrutiny. Once data is entered into the blockchain, it becomes immutable, ensuring the information's accuracy and integrity. This technology enhances security by distributing network data across multiple computers instead of relying on a single repository. By doing so, it also exposes fewer vulnerabilities to potential hackers. Furthermore, blockchain provides a fertile ground for the establishment of new businesses and competition with established industries [1–3].

Marketers in the pharmaceutical sector can effectively monitor product usage through the application of blockchain technology. With its capacity for drug tracing, blockchain technology is poised to significantly contribute to the pharmaceutical and healthcare industries by combating counterfeit drugs. It plays a vital role in pinpointing the origin of falsified medications. Additionally, blockchain ensures the security and privacy of patient records and, if implemented, can securely archive medical histories in a permanent record. Hospitals and their standard hardware infrastructure can leverage this decentralized network. Furthermore, researchers can make cost estimations for treatments, medications, and cures related to various illnesses and disorders using the data stored within this system [4, 5].

Blockchain functions as a decentralized network of ledgers that consistently appends new entries without ever modifying existing ones without unanimous agreement. The worth of a blockchain hash relies on the cryptographic link that binds each data segment to a history of recently updated data blocks. The distributed architecture of the blockchain ledger allows data processing without reliance on a central entity, resulting in transparency and accountability for all participants within the network. This decentralized approach reinforces and safeguards the system, rendering it resilient to single-point attacks. Patients can monitor the whereabouts of their medical records by storing them on a blockchain [6, 7].

1.1   Major Objective

With the aid of this technology, researchers can examine a tremendous amount of hitherto unexplored data regarding a particular group of people. The proper support for longitudinal investigations contributes to the advancement of precision medicine. We employ blockchain for healthcare realm instantaneously to lay up and bring up-to-date critical unwearied information like blood pressure and sugar levels with the assistance of the wearable expertise and Internet of Things (IoT). It enables medical professionals to monitor patients who are at high jeopardy and, in an emergency, counsel and notify the patient's loved ones and employers. Blockchain technology's decentralized architecture makes hacking safe and guards against conceding any one copy of the records [8, 9]. In this chapter, the following queries are addressed:

RQ1: Research using bags for blockchain technology in hard healthcare work

RQ2: To determine how blockchain technology might help the global healthcare culture

RQ3: To determine and talk about blockchain technology's facilitators for stimulating healthcare armed forces

RQ4: To establish a "Unified Work-Flow Process" for the implementation of blockchain technology in healthcare service delivery

RQ5: To discover, in addition to talking about essential blockchain applications in healthcare

Using quantum computers in healthcare can change how diseases are identified and treated. We can execute calculations that are not conceivable with classical computers by exploiting the capabilities of quantum computers. This might result in innovations like disease detection, personalized treatment, and drug development. Although quantum computing in healthcare is stagnant in its infancy, it has enormous potential to transform the field.

1.2   Organization of Work

This chapter organizes in such a way that total blockchain technology and quantum technology cover applications in healthcare systems. This chapter starts with concepts of blockchain, followed by techniques for artificial neural networks used in healthcare systems. After that, other applications, issues, and challenges are discussed. Limitations are discussed before illustrating the future directions and conclusions.

2   Quantum Computers in the Medical Field

The medical sector is where quantum physics and technology have surprised society the most. Quantum sensors and quantum computers are only two examples of cutting-edge technologies that have the potential to improve our healthcare system significantly. Quantum technology ought to be commercialized in the same way as our healthcare system. Examples of quantum technologies and quantum physics in the healthcare realm include artificial intelligence, nanotechnology, and quantum computing. These strategies have significantly impacted the healthcare sector and will help it develop further [10].

3   Nanotechnology and Quantum Sensing in Healthcare

Quantum technology works so well with our healthcare systems because many biological processes occur in an environment with microscopic characteristics akin to quantum interactions; therefore, using quantum technology makes sense.

Nanotechnology, a branch of quantum technology, has profoundly impacted medicine. Many companies are using nanotechnology to make quantum sensors. These sensors will be able to obtain more accurate measurements for special medical equipment. A quantum technology start-up named MacQSimal seeks to replace the bulky, expensive MEG (magnetoencephalography) equipment. MacQSimal is suggesting the deployment of a helmet built of quantum sensors to deliver more accurate brain scans [11]. Other companies, including Metabolism Project, use diamond-made quantum sensors to create MRI cooling system alternatives. Other quantum sensors are anticipated to enhance disease diagnostics and enable earlier detection of cancer, Alzheimer's disease, or dementia. These cutting-edge methods would significantly improve patient quality of life while reducing healthcare expenses [11].

4   Quantum Computing and Drug Testing

Quantum computing is another way that quantum technology has influenced medical treatment. Due to internal quantum interactions, quantum computers can store high-quality information, unlike conventional computers or supercomputers. Quantum bit is utilized by a quantum computer (q-bit) in a superposition of 0 and 1 simultaneously, as opposed to a conventional computer, which uses a bit (a unit of information) that is either a 0 or a 1. Through the mechanism of superposition, two particles can both be in the same quantum state simultaneously. Regarding quantum computing, things get pretty sophisticated quickly, but superposition allows these computers to store considerably more data per unit volume and do complex mathematical calculations and algorithms much faster. Our healthcare systems have already started to include quantum computers. Partnerships between hospitals and quantum technology firms are starting to form, indicating strong growth for this segment of the quantum market [12]. The Cleveland Clinic and IBM have a partnership that is one of these. Longtime collaborators IBM and the Cleveland Clinic have teamed up to create the Discovery Accelerator, a facility that will leverage quantum technology, including quantum computing, to improve medical and life science breakthroughs. Due to this collaboration, drug trials will be completed more quickly and effectively, and diseases will be detected.

A quantum computer could speed up drug research and testing in the medical profession. Quantum computers could perform accurate simulations of a revolutionary treatment on virtual human patients in just a few hours. This would reduce the number of test participants for a study—human or animal test subjects—and save drug companies time and money. The business InSilico Medicine has already tested this technique; using a simulated algorithm, it created a new medication candidate in just 46 days. The invention and testing of novel drugs can be accelerated with quantum computers, potentially saving thousands of lives. This would spare drug corporations years of drug testing and medication development and potentially save them hundreds, if not millions, of dollars [13].

5 Additional Quantum Computer Applications in Healthcare

The safest radiation therapy, which would best target only the cancerous tissue and not healthy tissue for individuals receiving it, might be imitated by a quantum computer. Due to its powerful and speedy computational capabilities, quantum computing may also be able to sequence or analyze complete human genomes far more swiftly than a regular computer. This speedier genetic study could have benefits, such as better genetic tests for genetic disorders and more precise drug screens. Quantum computing further offers more secure medical data through quantum data encryption (which will be explored in another post) [14].

6 Applications of Quantum Computing in Health

Quantum computing is already a reality. Yes, it is still in its infancy, but constantly opening new doors for the health industry. With quantum computing, issues that would take years can now be resolved in seconds [10].

How? For instance, it is anticipated that further sophisticated simulated cleverness algorithms have been developed within the prospect, which will aid in searching for more effective treatment options based on the patient type or in studying the structures of complicated molecules. This will therefore make it possible to find novel medicines and materials for use in clinical settings because of the enormous

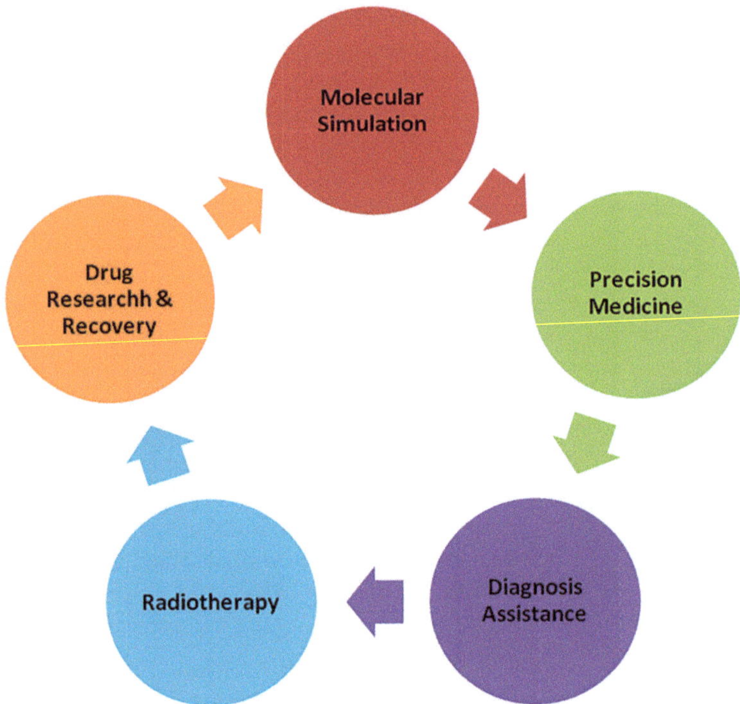

Fig. 1 Applications of quantum computing in health

flow of data and medical images that artificial intelligence and machine learning will generate. Moreover, as a result, medical data will be better protected in terms of security. However, there is much more. Indeed, the following are the **three main areas** in which **quantum computing causes a great revolution:**

According to recent studies, quantum computing has a distinct edge over traditional computing systems. The diagnosis and treatment of diseases are sped up gradually, thanks to quantum computing, which in some applications can dramatically cut computation times as of time to act. It inspires novel approaches to achieving a superior point of ability intended for convinced household tasks, new architectures, and methods, as depicted in Fig. 1.

6.1 Drug Rehabilitation

On the one hand, this technology allows for the quicker and more precise design of medicinal treatments. Quantum computing has the possibility to modernize drug treatment by enabling faster and more accurate simulations of molecular interactions. With the ability to perform complex calculations at an unprecedented speed, quantum computers can efficiently model the behavior of large molecules, such as proteins and enzymes, and predict their interaction with drug compounds. This predictive modeling allows researchers to design more effective drugs with fewer side effects. Additionally, quantum computing can aid in identifying novel drug targets, which may lead to the development of entirely new classes of drugs. Quantum computing can also assist in the optimization of drug manufacturing processes. By simulating chemical reactions at the molecular level, quantum computers can identify the optimal conditions for drug synthesis, leading to faster and more efficient drug production.

6.2 Recognition

Contrarily advances in quantum computing have made it possible to build techniques that can evaluate MRI tissues by comparing them to a vast number of previously stored tissues, detect the efficacy of chemotherapy after a single dose, or ascertain the success of treatment in minutes or days.

6.3 Data Management

Last but not least, quantum computers also power big data and AI, making it simpler to capture, sift, and analyze enormous volumes of complex data and uncover patterns in it in only a few seconds, significantly improving health data management. It will significantly speed up hygienic procedures.

By using blockchain technology to power a health information exchange, interoperability may finally reach its full potential. Blockchain-based schemes can do away with recent intermediaries and their costs. Blockchain holds great promise for all parties involved in the healthcare realm. This technology can bring disparate systems together and generate insights that will aid in determining the value of care. A national blockchain network for electronic health records (EHRs) could improve patient health outcomes and productivity. Figure 2 illustrates quantum computing and its applications in medicine.

6.4 Irradiation

A common approach to cancer treatment is radiotherapy, which involves the use of radiation to eliminate or inhibit the growth of malignant cells. Developing a radiation strategy is essential to minimize the radiation's impact on healthy tissues and organs, a process that involves addressing complex optimization challenges encompassing numerous variables. Consequently, it often requires multiple simulations to

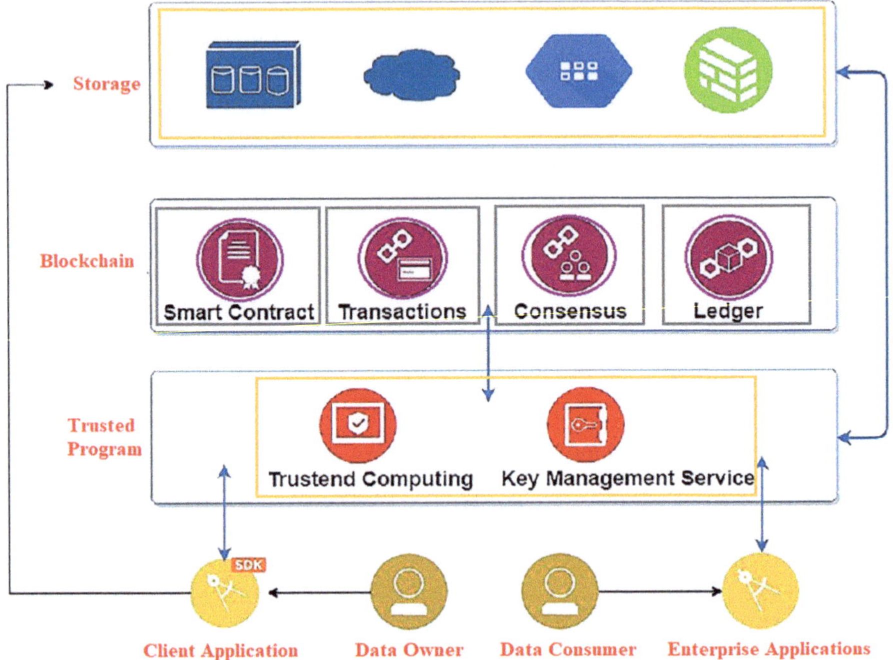

Fig. 2 Data handling innovations using blockchain

identify the optimal radiation plan. Quantum computing, on the other hand, allows for the exploration of a broad spectrum of possibilities between each simulation, enabling healthcare professionals to conduct multiple simulations simultaneously and generate the most effective treatment plan.

6.5 Drug Research and Interactions

The initial and pivotal phase in the design and discovery of a medication involves molecular comparison. Currently, organizations can employ traditional computers to conduct millions of such comparisons. Nevertheless, the capacity of conventional computers to compute molecules is constrained, particularly with respect to the size of molecules.

Quantum computing can therefore be used to compare more giant molecules. As a result, it will open the door for additional pharmacological developments and disease solutions.

In addition, quantum computing enables scientists in the medical field to impersonate complex molecular relations right down to the atomic level. It will therefore be essential for creating new medications and medical research. This will enable experts to simulate each of the 20,000 proteins found in the human genome shortly.

Additionally, simulations of interactions with models of new and current drugs will be used.

6.6　Healthcare Data

Patients expect their medical and healthcare information to be protected and secure. Therefore, it is essential to study and assess each hacking approach.

For instance, the company ID Quantique secures data using elements of quantum mechanics. As a result, one of the more practical uses is leveraging quantum entanglement, and quantum cryptography protects the data.

6.7　Genomics

The study of an organism's entire genetic makeup is called genomics. In other words, it includes bioinformatics, DNA sequencing techniques, and recombinant DNA. Additionally, it demands the assembly, sequencing, and analysis of the genomes' architecture and functions.

The most current methods also entail infringement of the DNA down into smaller portions. It also entails looking for specific biomarker subtypes and any disease-related alterations.

As a result, two critical consequences must be addressed. First of all, the procedure should be shorter. Second, when only manual operations are involved, it becomes slower. Traditional computers need to be more active in their tasks.

As a result, quantum computing is the best course of action because it has higher processing and storage power. Additionally, the results will be more precise, providing accurate diagnoses and customized medications.

Furthermore, it will enable experts to establish a comprehensive genome database for the identification of previously undiscovered biomarkers and mutations. The inclusion of various factors such as environmental and lifestyle considerations will also transform the approach to treatment.

6.8　Improving Imaging Solutions

Quantum imaging devices play a crucial role in generating precise images that enable the visualization of individual molecules. Furthermore, machine learning and quantum computing aid professionals such as doctors in interpreting these findings.

Additionally, quantum computing offers the interpretation of the data and its therapies, while machine learning aids in identifying bodily anomalies.

However, conventional MRIs can identify light and dark tissue patches, and the radiologist must assess these areas. In order to enable more accurate imaging, quantum imaging techniques help distinguish between distinct tissue types.

6.9 Accuracy Medication

Accuracy medicine is a field that focuses on offering individuals healthcare demands preventative and treatment techniques. In the future, personalized medicine that goes above and beyond conventional medical treatments will be necessary to suit the difficulty of the human being genetic scheme. Using EHRs, traditional ML has demonstrated success in identifying disease risks in the future. However, owing to superiority and sound, feature size, and the intricacy of relationships between features, employing traditional ML algorithms still has limitations. The fundamental way that quantum-enhanced ML could speed up medical advancements is by enabling models for drug inference.

6.10 Medicine Research and Invention

Medical professionals may simulate atomic-level molecular communications with quantum computing, which is necessary for medicinal investigation. This will be especially important for analytics, drug discovery, diagnosis, and therapy. Tens of thousands of proteins can now be encoded to simulate their interactions with medications, which was previously impossible, thanks to developments in quantum computing. Unlike traditional computer methods, quantum computing aids in processing this information orders of extent more efficiently. Doctors may evaluate massive data sets and variants using quantum computing to find the best trends. Gold nanoparticles can now be used to detect disease-specific biomarkers in the blood using established techniques like the bio-barcode test.

7 Blockchain

Fundamentally, blockchain is a distributed scheme for storing and retrieving transaction records. Specifically, a blockchain is a digital ledger containing shared, permanent evidence of peer-to-peer communication. It is constructed from linked transaction blocks. Blockchain relies on well-established cryptographic techniques to enable network interactions without the need for mutual trust in advance (such as storing, exchanging, and viewing information). In a blockchain system, all users preserve and distribute transaction records; there is no central authority. While keeping an immutable audit trail of all transactions, the blockchain enables trustless collaboration among network users. All participants in interactions with the blockchain become aware of them, and information must first be added after network verification.

7.1 *Blockchain Concept*

- A blockchain system facilitates peer-to-peer value transfers via computer consensus, eliminating the need for intermediaries. This is accomplished by utilizing a P2P network of computers running the protocol, each maintaining a duplicate copy of the transaction ledger, which operates on top of the Internet. Various forms of blockchain technology exist, including public, private, hybrid, and consortium blockchains. The specific advantages and drawbacks of each network significantly influence the most suitable application of a given blockchain network.
- Since the initial type of blockchain technology, the public blockchain, was used to create Bitcoin and other cryptocurrencies, they advanced distributed ledger technology (DLT). Centralization now has drawbacks, such as a need for more security and transparency. Instead of retaining the data in one location, DLT disseminates it around a P2P network. It becomes essential for some form of data authentication due to decentralization.
- A private blockchain network operates within a restricted environment, like a closed network or under the control of a single entity. It shares similarities with a public blockchain network in terms of peer-to-peer connectivity and decentralization but is considerably smaller in scale. In a private blockchain, the network's creator always has knowledge of the users, which is in contrast to public blockchain networks where users on the open Internet remain fully anonymous. This difference hinders the development of permissionless solutions.
- Hybrid blockchain, a type of blockchain that incorporates features from both private and public blockchains, is occasionally adopted by organizations seeking to benefit from both approaches. This allows businesses to establish a private, permission-based system alongside a public, permissionless system, offering control over access to specific blockchain data and determining which data is made publicly available.

8 The Rationale of the Study

The pace of progress in the healthcare industry continues to accelerate, creating a pressing need for top-tier medical facilities supported by cutting-edge technology. Blockchain has the potential to bring about substantial changes in how healthcare operates. Furthermore, the healthcare landscape is evolving in favor of a patient-centered approach, emphasizing two critical elements: sustained access to essential healthcare resources. Thanks to blockchain technology, healthcare providers can offer exceptional patient care and services. The time-consuming and repetitive process of sharing health information, a significant contributor to high healthcare costs, can now be swiftly addressed with technology. The general public can actively participate in health research initiatives through blockchain technology. Enhanced public health research and data sharing will enhance the treatment of many individuals [13, 15, 16].

Historically, constraints related to interoperability, privacy, and data sharing have been the primary challenges in population health management. Blockchain technology proves to be a dependable solution for this specific issue. When effectively implemented, this technology enhances security, data exchange, interoperability, integrity, real-time updates, and access. Data security is a major concern, especially in wearable technology and personalized care. Blockchain technology addresses these challenges by ensuring that patients and medical professionals can collect, transmit, and consult secure data across networks without compromising security concerns [17, 18].

8.1 Blockchain the Same as an Enabler of Countrywide Interoperability

A recently unveiled collaborative roadmap by the Office of the National Coordinator for Health Information Technology encompasses critical technological and legal prerequisites, such as the following:

1. A widespread, safe network infrastructure
2. Verifiable participant identification and authentication
3. Consistent portrayal of, among other things, access permissions to electronic health records

However, current technologies can only partially satisfy these requirements due to problems with security, privacy, and ecosystem interoperability.

9   Innovations

We point out numerous types of quantum computers, such as one-qubit computers [19], two-qubit computers [20], and higher-qubit quantum computers. The first five-qubit quantum computer was created in 2000, which led to significant advances in the field [21]. Since then, other significant developments have been made, with IBM's most recent quantum computing chip, which has 128 qubits, being the most well-known quantum computer of the modern age [22]. However, according to the literature, 50 qubits are required to achieve quantum supremacy [23].

Table 1 Association of classical computing vs. quantum computing

	Classical computing	Quantum computing
Computing unit	Calculate with transistors which can take two levels, 0 and 1	Calculate with qubits that can represent both 0 and 1 simultaneously
Computing facility	Capability increased linearly (1:1) with the number of transistors	Capability increased exponentially with the number of qubits
Error rates and environment	Low rates of error. Able to function at room temperature	High rates of error. Must be kept extremely cold
suitability	Suitable for routine (non-compute-intensive) processing	Suitable for complex(compute-intensive) processing

The advantages, drawbacks, and practicalities of the classical and quantum computing paradigms are compared in Table 1. In divergence to conventional computers that drive in terms of bits, quantum computers operate employing quantum bits, or "qubits," which have two states and can simultaneously denote one bit in the states of "1" and "0." Quantum physical systems that use an electron's spin and a photon's direction enable qubits.

The inability to determine the precise location of a spinning electron at any one moment is known as quantum superposition. Instead, a probability distribution is used to estimate the superiority of the electron being present everywhere at all times. Quantum computers, which use a set of qubits for calculations and hence speed up processing, are ticked by superposition. The processing power of a q-bit quantum computer increases exponentially in the form of 2q since a qubit can exist in two states. The contradictory fact that an entangled pair of electrons always spins in the opposite direction and impacts one another through time and space even when they are not physically connected is the quantum entanglement feature, which Einstein called "spooky." Numerous fields, including communication, image processing, information theory, electronics, and cryptography, can benefit from quantum computing. As quantum computers become more accessible, usable quantum algorithms are starting to appear. The concept that a fully developed quantum paradigm will address various computing issues has also been bolstered by the continuous efforts to design physically scalable quantum computing equipment. The current processing capacity can only solve a few computing issues.

10　Techniques for Artificial Neural Networks Used in the Healthcare System Using Blockchain

The blockchain algorithms powered by artificial intelligence ensure secure verification of PHR data from medical institutions and precise corroboration of medicinal records as presented weakness. Artificial intelligence has recently gained popularity, thanks to the Fourth Industrial Revolution, and stimulated research into numerous technologies. This is crucial to both healthcare and the status of the healthcare sector. Additionally, blockchain is highly secure because it verifies and encrypts these medical data in case it is breached or leaked. This paper highlights the issues with artificial intelligence blockchains and acknowledges issues with blockchain, AI, neural networks, healthcare, etc. These issues are why EHR solutions still need to be implemented. When these EHRs are operational, and data transmission and verification between hospitals are finished, it will be simpler to ensure isolation in the future. Several verification and performance evaluation metrics have also been set to collect the TPS of medical data and carry out standardization work in the future.

10.1 Applications of Blockchain in Healthcare

Due to its data management capabilities in terms of decentralization and security using distributed data technology, blockchain has caught the markets' interest. Various sectors are currently appreciating its reimbursement and recognizing the impending of blockchain in their respective fields due to its comparatively successful deployment in the financial industry. The reimbursements of blockchain knowledge are described in the healthcare segment. Blockchain technology offers numerous prospects for the healthcare industry to enhance the confidentiality and interoperability of patient health data. This article talks about several important uses in healthcare.

10.2 Blockchain for Electronic Health Records (EHRs)

Maintaining a lifetime medical record and making it accessible to doctors, lab techs, and the timely care team is the ideal scenario for EHRs. In the current EHR deployment, an unwearied record is held in several organizations during the patient's lifetime. Doctors keep track of patient medical information, such as Ofc's issues, and often act out long after treatment. A patient manages a medical record in a blockchain implementation and is given "smart contract" based access to clinicians, as shown in Fig. 3.

10.3 Clinical Research and Blockchain

Although there has been significant success in affecting checkup accounts as of document to digital proceedings, the healthcare sector is at rest functioning throughout troubles using patient data exchange between organizations and providers. Patients may visit different cities during their lives, switch health insurance policies, and seek treatment from different healthcare professionals. Most of their thorough medical records are kept at a private hospital or provider. Additionally, each medical department must save patient data using its semantics and storage format. Data sharing results as a result.

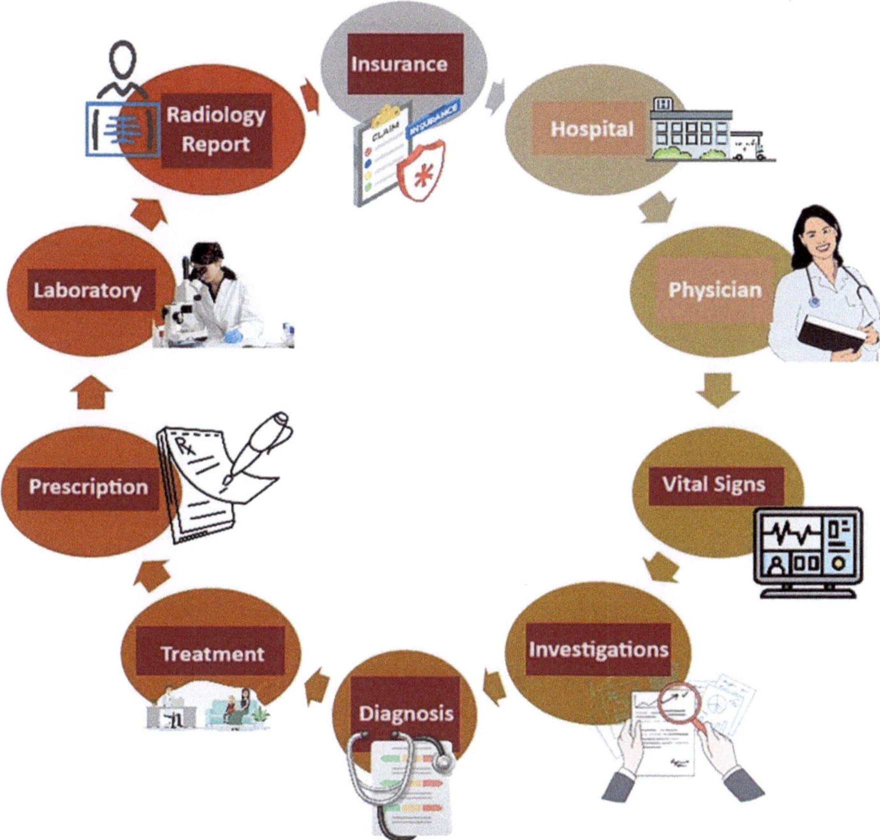

Fig. 3 Cryptocurrency for electronic health records

These obstacles are partially caused by the data's nature (protected health information) or by the fact that the information was prohibited during an exchange. Using real-world data to examine and evaluate their hypotheses in realistic settings is beneficial in treating patients in many institutions and clinical research.

10.4 Drug Supply Management and Blockchain

The supply chain management's transparency and security are among the industry's biggest challenges. Every year, drug losses cost the pharmaceutical industry millions. The effectiveness and safety of the product are crucial since pharmaceutical company contract with possessions that honestly affect the life of their customers. The product goes through several stages, from the manufacturer to the consumer, including transportation, maintenance, storage, redistribution, and sale. At this point, everything from a simple human error to nefarious intent can go wrong. This

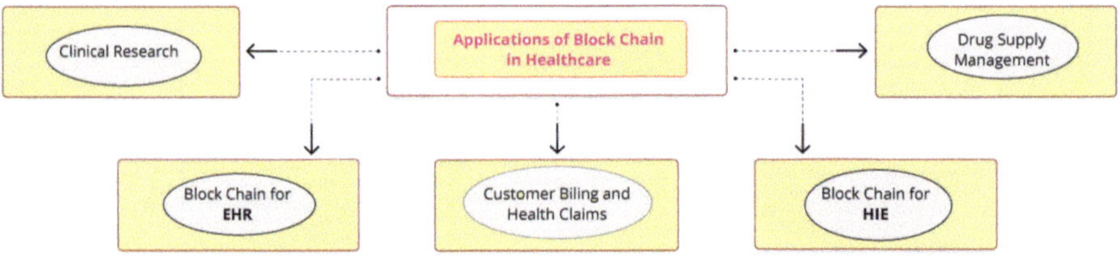

Fig. 4 Blockchain drug supply management

issue is challenging to pinpoint in the conventional system since supply chain participants typically can keep their records and exchange their knowledge. Additionally, these accounts have disadvantages if they are paper-based. These factors impede research efforts to find issues in the factor supply chain.

By enabling a shared supply chain among all supply chain participants, blockchain aids in resolving supply chain issues. The records on the blockchain are permanent, decentralized, and immutable at every point of the supply chain, as shown in Fig. 4. This removes the chance of introducing mistakes or fraud.

10.5 Blockchain in Healthcare and Consumer Billing

Fraudulent claims and billing are dangers to healthcare that should be removed and prevented. In the healthcare industry, fraud involving medical billing is still highly prevalent. The most prevalent types of healthcare fraud involve providers who overcharge for genuine services, claim fees for nonperforming services, and maintain services. The most frequent types of fraud are misrepresenting non-covered medical care for inpatient medical conditions, receiving claim money, and suffering financial losses.

In order to ensure that claims are processed quickly, save administrative expenses for providers and payers, and verify and delay claim information, numerous parties are involved. In the traditional claim process, communication between the parties is restricted behind walls. When processing payments and settling disputes, blockchain technology aids in reducing these difficulties. By automating the necessary workflow, blockchain technology enables all parties to share a single copy of contracts and billing data.

10.6 Blockchain Application in Hypoxic-Ischemic Encephalopathy

The primary purpose of hypoxic-ischemic encephalopathy is to provide a well-organized and secure deliverance mechanism by disseminating healthcare data across institutional and geographic borders. There are numerous aspects of a universal partnership strategy that must be taken into account:

- Infrastructure. A centralized data source is traditionally needed for data sharing, which raises the security risk footprint and forces reliance on a single centralized authority. Data privacy, failure to secure patient data, and potential financial and legal repercussions are all of the utmost significance.
- Differentiation. The information must be shared by all necessary parties so that they may comprehend its structure and significance.

All data on the blockchain is encrypted using each user's private key. By limiting the number of nodes or participants who have access to the data, the sending participant's public key preserves the data's privacy. Data from different network users is saved on their storage devices and occasionally sent to the HIE in typical HIEs. In this centralized storage system, updates to several patients' medical information could be jeopardized in the event of a storage failure as the amount of data per patient increases from various sources, including Webbers, doctors, and lab results.

The blockchain technology architecture enables each patient access to their data, unlike conventional systems where the central authority accesses and distributes data throughout the network. Access to medical records is only given to a few healthcare institutions (individuals or organizations). Real-time changes to all healthcare organizations are made possible by data shared over the blockchain network. It is possible to have safe access to the patient's 360-degree data via a distributed ledger. Due to everyone having access to an identical copy of the data, data duplication is minimized [24].

11 Healthcare Uses the Technology of Blockchain

Blockchain technology uses are described below and also shown in Fig. 5.

11.1 Securing Patient Data

Data and patient privacy protection have long been a top priority for the healthcare sector. Both consumers and regulators have requested certain safeguards over personal information. However, there have been numerous data breaches involving private health information.

Due to its immutable, decentralized, and completely transparent nature, blockchain stands away from home as a potential explanation for healthcare information safety. Each time someone interacts with data, a permanent record is created. This extreme transparency does not mean people's privacy worries are kept private. Transparency exists in logs, records, and transactions. However, with cryptography, individual identities are kept secret. Using blockchain in healthcare allows for the privacy of patients' identities and medical information while ensuring system security, as shown in Fig. 6.

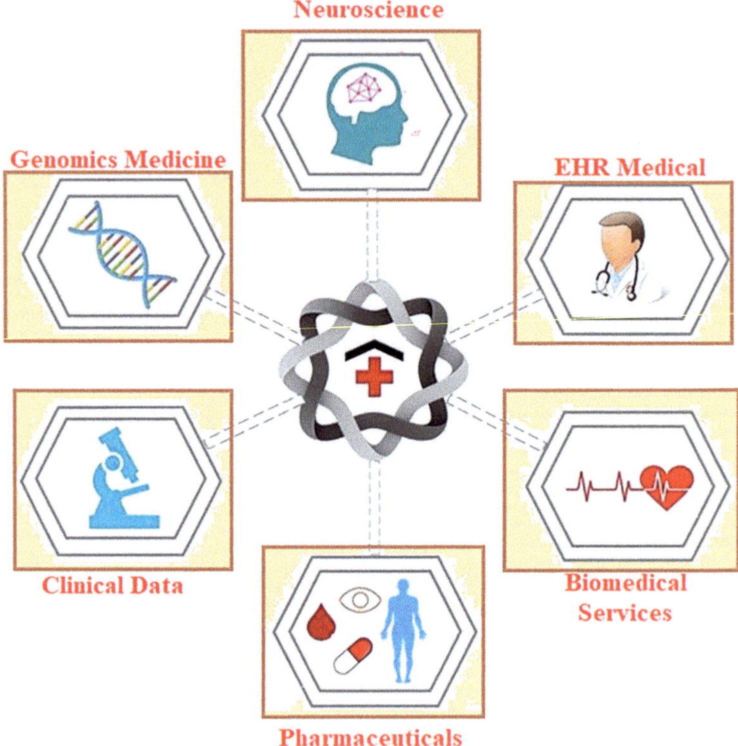

Fig. 5 Blockchain uses in healthcare

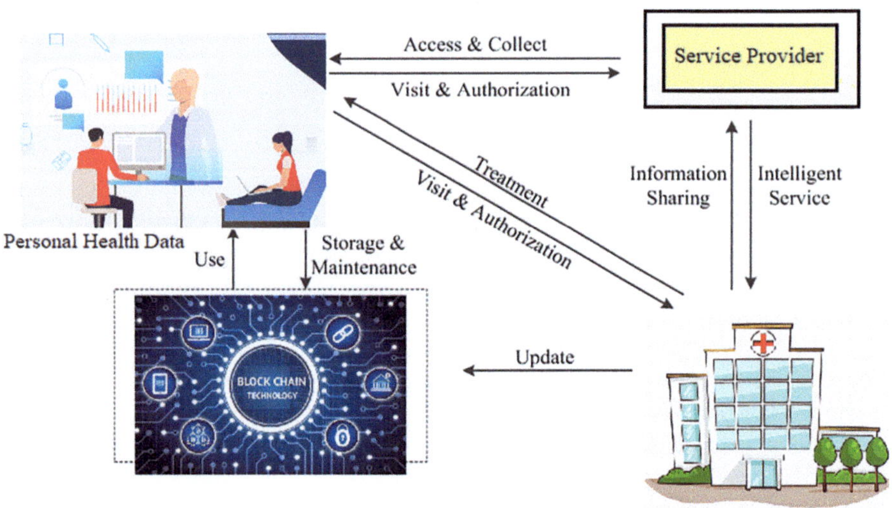

Fig. 6 Securing patient data

11.2 Supply Chain Management

The healthcare industry's supply chain needs to be more cohesive and intricate. From medicine development to consumer delivery, there needs to be more management control. The entire logistics chain is put under further stress by strict restrictions. Measurement and enforcement of quality control and compliance are difficult. Blockchain-based systems can help the entire industry with some of these logistical problems. These immutable, irrevocable transactions on the blockchain can help producers, for instance, track ingredients and guarantee levels of compliance. Blockchain technology's immutability can help with the compliance and quality control of healthcare supply chains and logistics.

11.3 Interoperability

Interoperability is one of the biggest problems with healthcare data management. Manufacturers, pharmacies, physicians, hospitals, and clinics are divided into independent organizations and systems. Each of these organizations manages its business and patient records through technology. As patients travel from one entity to another, medical records must be accessible.

This integration can benefit from blockchain technology. A blockchain ledger is spread by its nature. That implies that all parties with permission can access digital transactions containing patient data. Patients who change doctors can effortlessly share their medical history by changing only one permission. Blockchain technologies reduce the time and money needed for data transformation and reconciliation. The decentralized network allows authorized parties to access the patient data they require.

11.4 Medical Staff Credentialing

For those looking for a new job, the healthcare sector is flourishing. Additionally, the sector is heavily controlled. Recruiters and employers must thoroughly verify the credentials that applicants claim to have. The lives and health of people are in danger. Hiring an unqualified candidate can have disastrous results. Unfortunately, it is frequently an expensive and time-consuming process.

The issue of rapidly and effectively confirming medical credentials is one that blockchain systems promise to solve. Valid medical qualifications are entered into the unchangeable ledger. When issued, new credentials are added to the list because all blockchain records are designed to be verifiable. Only authorized users can add permission to see or amend the entries. The administration and verification of

Fig. 7 Medical personnel credentialing

credentials are quick and secure. Verifying medical credentials in an effective and fast manner is made possible by blockchain, as shown in Fig. 7.

11.5 Health Information Exchanges and Research Data

The most recent advancements in artificial intelligence, machine learning, and digital transformation have increased the value and usefulness of data. The healthcare industry is no exception and already has access to enormous amounts of medical data about patients, pharmaceuticals, clinical studies, and other topics. Deidentified data is crucial to investigate more potent drugs, cures, and projects.

Patients can find the security and anonymity they need with blockchain networks. Additionally, they ensure that the data that researchers and pharmaceutical companies need may be accessed continuously. This collaboration will make significant new medical discoveries and understandings that have the potential to completely change the way the world provides healthcare. Implementing blockchain technology could result in successful data-sharing networks that protect users' privacy.

11.6   Creating a Customized Healthcare Solution Using Blockchain Technology

Blockchain is a cutting-edge new technology that does not have to be mysterious or out of an organization's reach. Additionally, some enterprise blockchain use cases need specialized technology solutions that are not entirely accessible to the general public. Businesses want to guarantee that confidential information is safeguarded and limited from unauthorized access.

Private blockchain networks provide all the advantages of public blockchains plus the benefit of being entirely controlled by an entity. One such answer is the graphene framework. This framework included every requirement needed to run a network and was created for customized blockchain implementations. A barebones implementation, data storage, full peer-to-peer networking, and the necessary cryptography features are all included in prepackaged graphene components.

Focus on your company's needs rather than complex technology issues by employing a framework like this. However, as it is merely a framework, it only has some of the business rules required to make it operate. Frameworks get you up and running fast, whereas bespoke development provides those additional crucial rules.

For the development of unique blockchain applications, Chetu offers its services. For both public and private networks, we deploy high-performance blockchain technology. Additionally, we create custom wallet software, Bitcoin mining equipment, and blockchain applications.

12   Other Applications

12.1   Health Records by Electronics

Managing electronic health proceedings is one of blockchain technology's most significant applications. They could enhance patient care, save lives, and cut expenses. Electronic health records do, however, provide several difficulties. One significant issue is data security. The central databases where electronic health records are now kept are susceptible to hackers and data breaches. It raises the possibility of patient data being compromised. Blockchain technology, which offers a secure, decentralized way to store electronic health records, can be used to solve this problem. Patients may access their medical records whenever they want, and blockchain technology ensures that only authorized people can see them. Many blockchain-based healthcare start-up companies are already developing electronic health record systems for their organizations [2].

12.2 Drug Traceability

The traceability of a drug product requires following it along the supply chain from manufacturer to patient. Most drug tracing programs now use a centralized database, which is prone to manipulation or hacking. According to the WHO, up to 30% of medicines can be fake in some countries in Asia, Africa, and Latin America. The blockchain system would offer a decentralized, unchangeable record of each movement of medication products. Regulators could then spot and recall fake goods [19].

12.3 IoT Security for Remote Monitoring

The Internet of Things (IoT) is primarily used in the healthcare industry to monitor patients remotely. Your blood pressure, body temperature, and heart rate can all be measured by wearable sensors and other Internet of Things (IoT) gadgets. This information is then sent to a hub for use in monitoring by medical professionals.

IoT devices are not always secure, making them vulnerable to hacking and data breaches. Hackers could steal patient data, which could cause severe concern in the healthcare industry. Blockchain technology is the best solution to this problem since it provides a secure method of storing and sharing data. A secure Internet of Things network can be created using blockchain technology. The IOTA platform, for instance, provides a secure IoT environment [25].

12.4 Medical Staff Credential Verification

Verifying the credentials of medical professionals is another crucial application. According to Statista, there are 1,073,616 licensed doctors in the country. With such a significant number, it can be challenging to confirm the credentials of every medical expert. An increasing number of businesses are using blockchain technology to validate the credentials of medical workers. Organizations can swiftly and securely check the credentials and experience of medical professionals, thanks to blockchain technology. Furthermore, blockchain technology offers a permanent record of credentials that may be utilized to track the professional histories of medical staff members [17].

12.5   Management of the Supply Chain

The delivery sequence in the healthcare sector needs to be simplified and more cohesive. From medicine development through consumer delivery, management has minimal control over the entire process. Heavy rules make it challenging to monitor and enforce quality control and compliance. Supply chain management problems are aided by blockchain technology. Immutability, ingredient tracking, and compliance are all made possible by it. Blockchain can be used to trace the source of fake medicines and technology [13].

12.6   Blood Plasma Supply Chain

Blockchain technology can improve plasma supply chains and give current information to all players in the healthcare sector. Plasma derivatives are needed for numerous lifesaving procedures, but the current supply chain is complicated. Plasma is gathered, processed, and dispersed by several hospitals and blood banks worldwide [26, 27].

Unfortunately, the system frequently has inconsistencies and inefficiencies, which results in critical plasma shortages when patients need them most. By offering a decentralized platform on which all parties can share information and follow plasma shipments in real time, blockchain thus offers a solution to these issues. Blockchain might guarantee that patients always have access to lifesaving treatments in this way.

12.7   24/7 Data Monitoring

Monitoring data is essential for managing and maximizing electronic health records in the healthcare industry. Patient information must be maintained and available around the clock, including blood pressure readings and medications from doctors. Unfortunately, the current healthcare infrastructure needs to have these capabilities due to interoperability problems, data breaches, and other problems.

Blockchain is helpful in this situation. Due to its distributed nature, it offers continuous data accessibility and enhances the organization of electronic healthiness proceedings. Additionally, blockchain contributes to resolving the interoperability problem by offering a uniform platform for data interchange across numerous healthcare providers.

13   Blockchain Technology Issues and Challenges in the Healthcare Sector

Blockchain technology has various potential benefits for the healthcare industry but has several implementation issues.

13.1   HIPAA Compliance and Data Privacy Regulations

Blockchain technology can help healthcare organizations comply with HIPAA. However, the query of how to employ blockchain technology in a way that complies with these rules still needs to be answered.

13.2   Data Immutability

Once the data is submitted, updates cannot be done using blockchain. Immutability limits scalability, while, on the one hand, increasing security is guaranteed. It would be necessary to modify the current design to allow for the recording of alterations or, in the short term, use on-chain and off-chain data partitioning to lecture to the immutability issue.

13.3   Scalability

The blockchain system's scalability is another drawback; because the amount of blocks grows, the scheme becomes slower and uses added computing resources. This may be more challenging in an IoT setting where data must flow instantly.

14   Implementation Challenges and Considerations

Although there are many prospects presented by blockchain technology for the healthcare realm, it still needs to be developed entirely and cannot be used as a quick fix. Before a healthcare blockchain is implemented by companies nationwide, several technological, organizational, and behavioral economics obstacles need to be solved [28, 29].

Blockchain's Limitations in Healthcare Management
Figure 8 illustrates the blockchain's limits in the management of healthcare.

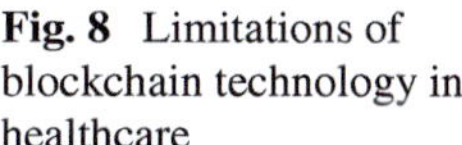

Fig. 8 Limitations of blockchain technology in healthcare

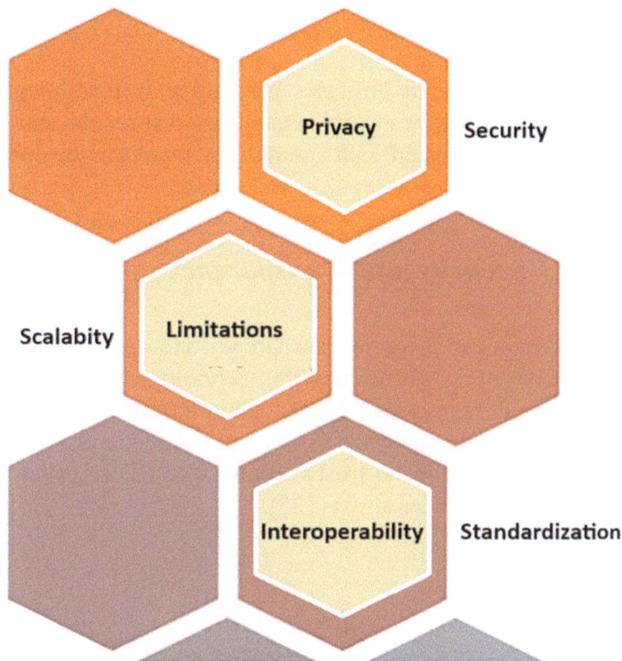

14.1   Cross-Chain Interoperability Issue

Most large enterprises with needs for large-scale data management are now moving toward adopting blockchain solutions. No unique platform is accessible, and no efficient protocols provide the necessary exclusivity. One of the problematic elements is interoperability concerns. Creating a distributed enabling ecosystem for healthcare requires the interoperability of cross-chain blockchain solutions. The limitation of node ability in size and gazette of scalability and protocols creates the healthcare application more dependable. Using a cross-chain approach improves communication with the fully linked network and reduces transaction costs. Direct connectivity across multiple remote healthcare chains helps interoperable communication within the ecosystem. Additionally, specific unresolved interoperability challenges in the healthcare industry remain, including transactional trust and the frequency of bottlenecks in serverless node transactions [30].

14.2   Scalability

Developers have addressed scalability by building new chains that link to the primary blockchain [31]. Off-chain solutions, often known as Layer 2, are what they are. The issue of scalability has specific challenges, including the maximum capacity of the block, high volume of transactions, blockchain transaction limits, and so on [10].

14.3 Latency

One of the biggest obstacles preventing further integration of blockchain is its latency issue. However, its outstanding test results show that Quick Node has primarily solved this difficult conundrum, thanks to its globally distributed nodes [31].

14.4 Security and Transparency

Blockchain can increase transparency and open up data. Because of the rise in the popularity of cryptocurrency, this technology is now more widely used than intended in financial transactions. Blockchain offers an expanding record of every transaction that has already happened in a domain, safeguarded against alteration and adulteration, allowing for the trustworthiness of the recorded information and real-time access to these transactions [25].

14.5 Lack of Standardization

There has been a recent exploration of standardization activity around blockchain and digital assets. The models work adequately for producers and consumers. The World Economic Forum and Global Blockchain Business Council are working to meet blockchain standards globally [24].

14.6 Accessibility

In terms of accessibility, blockchain provides availability by allowing people with access to it to access information that has been encrypted. Since there are backups on numerous network nodes, this information is always accessible, preventing any information loss. Additionally, portability is made possible by blockchain, allowing the adoption of multiple programming languages and access through diverse sources. However, its high cost and energy consumption may make it challenging to deploy, reducing its availability. Additionally, blockchain does not promote publicity; documentation is only considered when building the infrastructure that will utilize this ledger [17, 18].

15   Conclusions and Future Directions

Blockchain technology has undergone a transformative evolution since its initial implementation in Bitcoin, emerging as a versatile technology with applications across various sectors, including healthcare. This report offers an ample assessment of the current state of blockchain technology's utilization in healthcare and healthcare security, highlighting recent advancements, applications, and associated challenges. The study's goals included identifying novel applications of blockchain technology in the healthcare sector and those that are already in existence, along with an exploration of the challenges and drawbacks associated with these applications and an examination of the methodologies employed to develop them.

Based on our research, blockchain technology offers multiple applications within the healthcare sector, including pharmaceutical, electronic medical records management, advancement of biomedical research and medical supply chain oversight, education, remote patient monitoring, and health data analysis, among others. Several prototypes of healthcare applications based on innovative blockchain concepts such as permissioned blockchain, off-chain storage, and smart contracts have been established. However, further study is essential to fully comprehend, define, and assess the possibility of blockchain technology in healthcare. Additional research is also required to provision ongoing initiatives intended at tackling the challenges associated with the integration of blockchain technology in the healthcare industry.

The usage of blockchain technology in other healthcare schemes with a similar structure and the fight against the COVID-19 disease have proven to be very beneficial. The most modern healthcare schemes integrate sensor data that can be employed to monitor patients while also preserving their anonymity and privacy. This technology enables users to access large amounts of data at any time by storing it organized and securely. This elevates it to a higher class than all other technologies, which demotes them. They can be utilized whenever and wherever necessary since they can be quickly and efficiently supplied whenever and wherever needed. Patients may be located and tracked down in minutes with the quantum blockchain technology and network. It is possible to hide data in a quantum blockchain while ensuring it is safe and easy to access. It is capable to process patient data more swiftly through retaining its reliability by employing quantum computing and blockchain technologies. There has been much speculation about quantum technology and blockchain technology. This inquiry aimed to ascertain whether either of these technologies had any active medical applications at the time. Numerous technologies, including blockchain, quantum physics, machine learning, ratification intelligence, and drones, are the subject of research in this area.

References

1. Khezr, S., Moniruzzaman, M., Yassine, A., & Benlamri, R. (2019). Blockchain technology in healthcare: A comprehensive review and directions for future research. *Applied Sciences, 9*(9), 1736.
2. Kumar, T., Ramani, V., Ahmad, I., Braeken, A., Harjula, E., & Ylianttila, M. (2018). Blockchain utilization in healthcare: Essential requirements and challenges. In *2018 IEEE 20th international conference on E-health networking, applications, and services (Healthcom)* (Vol. 17, pp. 1–7). IEEE.
3. Moona, G., Jewariya, M., & Sharma, R. (2019). Relevance of dimensional metrology in manufacturing industries. *MAPAN, 34*, 97–104. https://doi.org/10.1007/s12647-018-0291-3
4. Kassab, M. H., DeFranco, J., Malas, T., Destefanis, G., Laplante, P., & Neto, V. V. (2019). Exploring research in blockchain for healthcare and a roadmap for the future. *IEEE Transactions on Emerging Topics in Computing, 9*, 1–1.
5. Shen, B., Guo, J., & Yang, Y. (2019). MedChain: Efficient healthcare data sharing via blockchain. *Applied Sciences, 9*(6), 1207.
6. Chelladurai, U., & Pandian, S. (2021). A novel blockchain-based electronic health record automation system for healthcare. *Journal of Ambient Intelligence and Humanized Computing, 13*(1), 693–703.
7. Zhang, P., Schmidt, D. C., White, J., & Lenz, G. (2018). Blockchain technology use cases in healthcare. In *Advances in computers* (Vol. 111, pp. 1–41). Elsevier.
8. Yaqoob, I., Salah, K., Jayaraman, R., & Al-Hammadi, Y. (2021). Blockchain for healthcare data management: Opportunities, challenges, and future recommendations. *Neural Computing and Applications*, 1–6.
9. Liang, X., Zhao, J., Shetty, S., Liu, J., & Li, D. (2017). Integrating blockchain for data sharing and collaboration in mobile healthcare applications. In *2017, IEEE 28th annual international symposium on personal, indoor, and mobile radio communications (PIMRC)* (pp. 1–5). IEEE.
10. Solenov, D., Brieler, J., & Scherrer, J. F. (2018). The potential of quantum computing and machine learning to advance clinical research and change the practice of medicine. *Journal of Missouri Medicine, 115*(5), 463–467.
11. Varshney, A., Garg, N., Nagla, K. S., et al. (2021). Challenges in sensors technology for industry 4.0 for futuristic metrological applications. *MAPAN, 36*, 215–226. https://doi.org/10.1007/s12647-021-00453-1
12. Holbl, M., Kompara, M., Kamisalic, A., & Zlatolas, L. N. (2018). A systematic review of the use of blockchain in healthcare. *Symmetry, 10*(10), 470.
13. Dhillon, V., Metcalf, D., & Hooper, M. (2021). Blockchain in healthcare. In *Blockchain-enabled applications* (pp. 201–220). Berkeley.
14. Kaushik, K., Ahuja, N. J., & Kumar, A. (2023). *Sustainable security practices using blockchain, quantum, and post-quantum technologies for real time applications* (Vol. 2023). Springer Nature. https://doi.org/10.13140/RG.2.2.17354.64961
15. Farouk, A., Alahmadi, A., Ghose, S., & Mashatan, A. (2020). Blockchain platform for industrial healthcare: Vision and future opportunities. *Computer Communications, 154*, 223–235.
16. Ekblaw, A., Azaria, A., Halamka, J. D., & Lippman, A. (2016). A case study for blockchain in healthcare: "MedRec" prototype for electronic health records and medical research data. In *Proceedings of IEEE Open & Big Data Conference* (Vol. 13, p. 13).
17. Dimitrov, D. V. (2019). Blockchain applications for healthcare data management. *Healthcare Informatics Research, 25*(1), 51.
18. Shen, B., Guo, J., & Yang, Y. (2019). MedChain: Efficient healthcare data sharing via blockchain. *Applied Sciences, 9*, 1207.
19. Hanneke, D., Home, J., Jost, J. D., Amini, J. M., Leibfried, D., & Wineland, D. J. (2010). Realization of a programmable two-qubit quantum processor. *Nature Physics, 6*(1), 13–16.

20. Balaganur, S. (2022). Man's race to quantum supremacy: The complete timeline. *Analytics India Magazine.* https://analyticsindiamag.com/race-quantum-supremacy-complete-timeline. Accessed 22 Feb 2022.

21. Ball, P., et al. (2021). First quantum computer to pack 100 qubits enters crowded race. *Nature, 599*(7886), 542–542.

22. Boixo, S., Isakov, S. V., Smelyanskiy, V. N., Babbush, R., Ding, N., Jiang, Z., Bremner, M. J., Martinis, J. M., & Neven, H. (2018). Characterizing quantum supremacy in near-term devices. *Nature Physics, 14*(6), 595–600.

23. Zhong, H.-S., Wang, H., Deng, Y.-H., Chen, M.-C., Peng, L.-C., Luo, Y.-H., Qin, J., Wu, D., Ding, X., Hu, Y., et al. (2020). Quantum computational advantage using photons. *Science, 370*(6523), 1460–1463.

24. Zhang, X., Poslad, S., & Ma, Z. (2018). Block-based access control for blockchain-based electronic medical records (EMRs) query in eHealth. In *Proceedings of the 2018 IEEE global communications conference (GLOBECOM), Abu Dhabi, UAE, 9–13 December 2018* (pp. 1–7).

25. Zhang, X., & Poslad, S. (2018). Blockchain support for flexible queries with granular access control to electronic medical records (EMR). In *Proceedings of the 2018 IEEE international conference on communications (ICC), Kansas City, MO, USA, 20–24 May 2018* (pp. 1–6).

26. Wang, H., & Song, Y. (2018). Secure cloud-based EHR system using attribute-based cryptosystem and blockchain. *Journal of Medical Systems, 42*, 152.

27. Sinitsyn, N. A. (2018). Computing with a single qubit faster than the computation quantum speed limit. *Physics Letters A, 382*(7), 477–481.

28. Al Omar, A., Bhuiyan, M. Z. A., Basu, A., Kiyomoto, S., & Rahman, M. S. (2019). Privacy-friendly platform for healthcare data in cloud based on blockchain environment. *Future Generation Computing Systems, 95*, 511–521.

29. Genestier, P., Zouarhi, S., Limeux, P., Excoffier, D., Prola, A., Sandon, S., & Temerson, J. M. (2017). Blockchain for consent management in the eHealth environment: A nugget for privacy and security challenges. *Journal of the International Society for Telemedicine and eHealth, 5*, GKR-e24.

30. Mazlan, A. A., Daud, S. M., Sam, S. M., Abas, H., Rasid, S. Z. A., & Yusof, M. F. (2020). Scalability challenges in healthcare blockchain system - A systematic review. *IEEE Access, 8*, 1–1. https://doi.org/10.1109/ACCESS.2020.2969230

31. Ksibi, A., Mhamdi, H., et al. (2023). Secure and fast emergency road healthcare service based on blockchain technology for smart cities. *Sustainability, 15*(7), 5748. https://doi.org/10.3390/su15075748

Blockchain Technology and Quantum Computing: A Promising Solution for the Healthcare Industry and COVID-19 Pandemic

Galiveeti Poornima, Deepak S. Sakkari, P. Karthikeyan, T. N. Manjunath, and K. Saritha

Abstract Blockchain technology could benefit both the healthcare industry and efforts to stop the COVID-19 pandemic. This study examines the significance of blockchain and concludes that, in the future, healthcare systems will use blockchain technology and associated protocols to automatically monitor patients, collect data from sensors, and store data securely. With this technology, it is possible to store a lot of data in different places and keep it safe. It also makes it possible to access the data whenever and wherever it is needed. As a result, the process of carrying out operations is much simpler when this technology is used. Quantum blockchain can maximize quantum computing's benefits, such as the ability for using infrared cameras and the speed at which sick people can be found and observed. This includes the capability of acquiring thermal imaging based on quantum computing. Another tool that can be employed to protect the data privacy of records while simultaneously ensuring their authenticity and making them easily accessible is known as the quantum blockchain. If an amalgamation of quantum computing and blockchain technology is employed to process medical records, there is the potential for a more private and expedited process compared to the current state. This paper examines

Deepak S. Sakkari, P. Karthikeyan, T. N. Manjunath and K. Saritha contributed equally with all other contributors.

G. Poornima (✉) · K. Saritha
School of CSE & IS, Presidency University, Bangalore, Karnataka, India
e-mail: galiveetipoornima@presidencyuniversity.in; kuppala.saritha@presidencyuniversity.in

D. S. Sakkari
Department of CSE, SKIT, Bangalore, Karnataka, India

P. Karthikeyan
Department of CS & IS, National Chung Cheng University, Chiayi, Taiwan

T. N. Manjunath
Department of ISE, BMSIT, Bangalore, Karnataka, India
e-mail: manju.tn@bmsit.in

S. Pulipeti et al. (eds.), *Quantum and Blockchain-based Next Generation Sustainable Computing*, Contributions to Environmental Sciences & Innovative Business Technology, https://doi.org/10.1007/978-3-031-58068-0_9

the potential advantages and blockchain applications and quantum computing within the fields of pharmacy, medicine, and healthcare systems. With this framework, the study assesses and contrasts quantum and blockchain technologies with other innovative information and communication technologies, including artificial intelligence, drones, machine learning, and similar innovations. Within this context, the terminology "blockchain technologies" is used to refer to both quantum- and blockchain-based technologies.

Keywords Blockchain · Data records · Information · Quantum computing · Quantum machine learning

1 Introduction to Quantum Computing

A promising method of computing called quantum computing (QC) is built around the strange occurrences of quantum physics. This is an excellent example of the harmonious integration of mathematics, physics, computing, and computational modeling. It outperforms conventional computer systems in terms of computational capacity, reduced energy consumption, and incredibly rapid speed by monitoring the interactions of extremely small particles like atoms, positrons, and photons. Given that quantum theory is a wider concept of scientific knowledge than standard mechanics, it makes a significant contribution to a more open-ended computing framework called quantum computing that can resolve issues that traditional computing cannot. QC uses its quantum bits, also referred to as "qubits," as opposed to conventional computers, which process and store data using binary data 0 and 1. Although the most amazing supercomputer currently available ends up taking thousands of years, QC can quickly break through the encryption used today. Although it is likely that QC will be able to breakdown some of the present encryption methods, it is also likely that they will come up with new ones that cannot be broken. Such small computers are unable to utilize transistors, logic gates, or integrated circuits. Therefore, atoms, neutrons, positrons, and ions are used as bits, along with information about how they spin and where they are. To create new combinations, they can be stacked. To be stronger, they can work in parallel and make good use of memory. The Church-Turing theorem is wrong because the QC computing paradigm makes better use of the systems than any other computing paradigm.

Quantum computing (QC) is characterized by three fundamental features, as detailed in [1]: entanglement, interference, and superposition. In quantum computing, superposition states to the capability of a quantum method to exist concurrently in two distinct locations or configurations. It differs significantly from its conventional counterparts, has binary restrictions, and permits impressive parallel processing at high speed. The QC system stores data twice. In conventional physics, wave interference is analogous to interference in quantum mechanics. Wave interference results from the collision of two waves in the same space. Imagine each wave facing the same direction. If so, it creates waves with the sum of the individual waves'

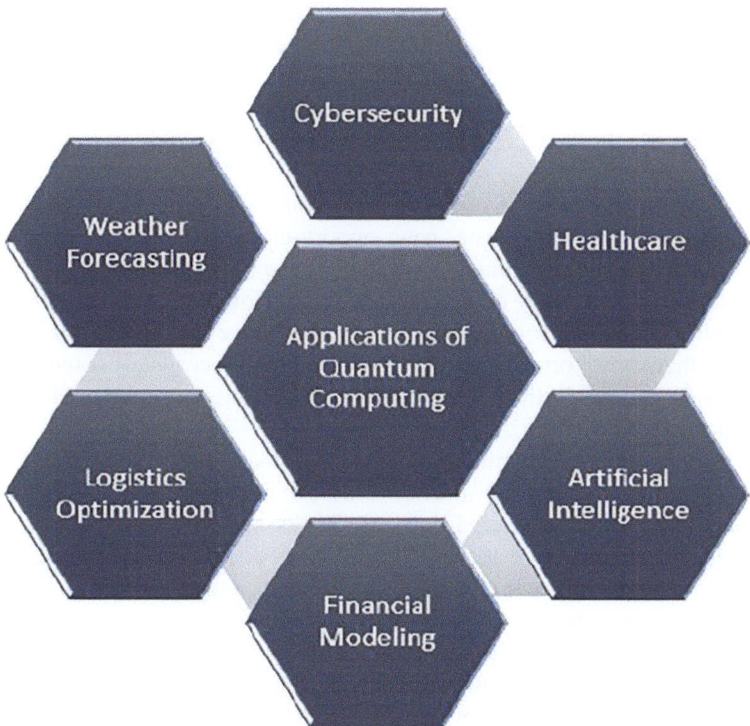

Fig. 1 Applications of QC

amplitudes (beneficial interference) or a wave without amplitudes (interferes destructively). Interference can increase or decrease the net wave. Quantum computers use entanglement. It makes reference to the manner in which two quantum bits and particles are connected. Even when kept apart by huge distances, such as when they are located at qubits, they have an exact and immediate relationship with the exact opposite ends of the universe. They are linked or share characteristics.

There are many uses for quantum computing, but Fig. 1 highlights some notable ones.[1] Cybersecurity [2], healthcare [3], artificial intelligence [4], financial modeling, weather forecasting, and logistics optimization are some of the main applications of quantum computing. The Internet security environment [5, 6] has become extremely vulnerable as a result of the increase in cyberattacks that occur around the world. Although businesses are implementing the necessary security structures, the process of using conventional digital computers has become difficult and unusable. Figure 1 illustrates the various uses of quantum computing, including some of the following:

---

[1] https://analyticsindiamag.com/top-applications-of-quantum-computing-everyone-should-know-about/

- *Healthcare.* Quantum computing is expected to offer significant benefits in healthcare that traditional computing cannot. Quantum computing requires new thinking, specialized skills, unique IT structures, and innovative business plans. Technology also impacts security. The healthcare sector is particularly concerned about security due to the commitments and challenges of the sector with respect to data privacy.
- *Cybersecurity.* Large-scale quantum computing will significantly increase computational power and create new opportunities to improve cybersecurity. Cybersecurity from the quantum era could have identified and thwarted attacks from this era before they could cause damage. However, it might turn out to be the ability of quantum computing to quickly solve the difficult mathematical puzzles that some forms of encryption incorporate because it could create new security holes. Institutions and other organizations can prepare for quantum cryptography now.
- *Financial modeling.* Quantum computing will likely be very beneficial to financial companies that can use it. They will have a greater ability, in particular, to evaluate large or unstructured datasets. Banks may improve their choices and provide better customer service by creating deals that are more pertinent. Quantum computers are promising when methods are driven by true data streams like share prices, and there is a lot of random noise.
- *Logistics optimization.* The logistics sector may take many advantages from quantum computing. Quantum computers would be used in conjunction with current CPUs to accelerate devices that use AI and machine learning. The logistics industry's route planning would be significantly impacted by quantum supercomputers. By examining all viable routing decisions and electing the most advantageous one attractive into account all factors, the leverage of quantum computing will indeed progress the use of warehouse modeling.
- *Artificial intelligence.* Both artificial intelligence (AI) and quantum computing are revolutionary technologies, and quantum computing is necessary for AI to advance significantly [7]. The processing power of traditional computers places limits on artificial intelligence, despite the latter's ability to run useful applications. Quantum computing could give artificial intelligence a boost in processing power, enabling it to tackle more difficult problems in a range of industrial and academic fields.
- *Weather forecasting.* Quantum computing has the potential to save lives and reduce annual property damage both on a local and global scale by providing weather forecasters with more sophisticated and accurate warnings of potentially devastating weather events. To learn about the state of computing and its expanding impact on a variety of industries, stay current with the 1QBit blog and our social media pages for news that goes beyond weather forecasting. Quantum computing can enhance weather forecasting and tracking, when subatomic optimization schemes are employed and vast volumes of data with many variables are handled quickly and effectively.

The most difficult QC tasks are the formulation and construction of pharmaceuticals. Drugs are typically developed through costly, risky, and time-consuming trial and error. Studies suggest that QC could be a useful tool to examine the effects of drugs on individuals, potentially saving pharmaceutical companies a significant amount of time and money. Since technological advances have permeated almost every aspect of modern life, artificial intelligence and machine learning are two of the most important issues at hand. Complex issues that would typically take years to resolve in traditional devices can be solved more quickly with QC. Accounting professionals manage enormous amounts of money, thus, even a slight modification to the anticipated return could have a significant effect. An additional use is algorithmic time to invest, which uses a computer to carry out complex operations to automatically start introducing share trades according to market circumstances. This is expedient, especially for large-scale transactions. A cutting-edge, effective algorithm called quantum annealing may one day outperform conventional systems. Ubiquitous QC can solve any computer engineering problem; nonetheless, it is not commercially obtainable.

The paper is structured into nine primary sections. The introduction to quantum computing is presented in Sect. 1, while Sect. 2 delves into the discussion of quantum Blockchain. Sections 3 and 4 shed light on innovative quantum-integrated technologies, including quantum drones, quantum AI, quantum satellites, and quantum machine learning, respectively. Section 3 offers an outline of the introduction of quantum blockchain in healthcare. Section 5 delves into photonic quantum computing for healthcare. Sections 6 and 7, respectively, accentuate quantum gates and circuits for healthcare and delve into quantum algorithms for healthcare. Sections 8 and 9 provide comprehensive insights into a comparative analysis of methods for healthcare based on quantum computing and the pros and cons of quantum blockchain for healthcare. Lastly, Sect. 10 concludes the paper and outlines the subsequent steps.

2   Quantum Blockchain

The blockchain technology is a distributed and decentralized hyperledger that copies every action or digital event and stores and distributes copies of them. Every transaction is confirmed by the large number of parties involved. Every transaction is recorded. Blockchain, a revolutionary technology that is transforming many industries, was unceremoniously introduced to the public with Bitcoin, its most developed version. Cryptocurrencies, such as Bitcoin, can be used to make purchases instead of using cash. Blockchain technology aided cryptocurrency development. Encoded, distributed, and decentralized QC and quantum information theory are based on quantum blockchain (QB). Once the information has been entered into the QB, it cannot be changed. More academics have focused on QB research in recent years, as QC and the quantum communication model have advanced. In the rapidly developing field of blockchains, QB Technologies has started an extensive research, development, and individual benefit program that includes virtual currency extraction and other sophisticated blockchain applications. To offer a fresh and cutting-edge viewpoint on blockchain technology, quantum will concentrate its research and development on cryptography, fusing the most complex implementation strategies and functional areas to quantum computing technologies, as well as deep learning AI.

With a market cap of $ 680 million last month, the cryptocurrency industry started to grow. Whatever happens after that, cryptocurrencies will have a bigger impact on the global financial system. Making sure that everyone uses digital money legally is the most difficult part of its implementation. In addition, it seems like the blockchain offers a promising substitute. By using cryptographic algorithms that are generally thought to be unbreakable, barring brute-force attacks, this ensures honesty. The authors of this paper [8] looked at recent developments in QB and briefly covered its advantages over conventional blockchain technology. The architecture and organization of the QB are discussed. Next, a portion of a larger blockchain is used to demonstrate how quantum technology can be applied there.

Li [8] shows that a crucial quantum advantage derives from the time entanglement rather than the spatial entanglement. All system components were experimentally realized. Additionally, their encoding method may alter the past. The article [9] investigates the fundamental issues, risks, and advantages of applying these technologies in the long term. We conclude our study with a summary of the important gaps in the field, methodological concerns, and recommendations for future research.

A distributed hyperledger protected from damaging modifications by cryptography is called a blockchain. Blockchain systems nevertheless promise an extensive variety of applications, despite relying on the SSL signature, which is vulnerable to outbreaks from quantum computers. Similarly, the cryptographic methods employed to create blocks also exhibit this characteristic, albeit to a lower degree. This suggests that quantum processing, which is closely linked to both sides of the tale, may possess an advantage in the pursuit of mining rewards. To solve the blockchain problem in the quantum era, researchers [10] present an experimental implementation of a public blockchain for the identification of cognitive people that uses the distribution of quantum keys over an urban fiber network. As part of this research, disciplines related to quantum mechanics will also be categorized according to how well each discipline is understood and studied.

In the paper [11], blockchain scheme and quantum cryptography are briefly introduced. The foundation and source of network-based communication techniques are discussed. This section provides a brief explanation of the concept of developing a p2p data transmission application that makes use of QB technology. The main goal of QB technology's data transmission is to establish long-lasting, reliable network connectivity.

3 Quantum Blockchain and Its Importance for Healthcare

The high level of complexity and expense of the healthcare ecosystem can be reduced with the help of healthcare datasets recorded digitally and insurance companies. General privacy laws allow data owners to identify why their information is stored and used. However, the Internet is an open network, and, as a result, healthcare data are exchanged over it; malicious activities such as stealing sensitive information or changing stored data are possible. As a result, conventional medical systems can find it difficult to maintain participant security and anonymity [12]. Blockchain has developed into a platform that increases the effectiveness of the existing healthcare system while protecting the confidentiality and security of all parties. In this article, motivated by their findings, the authors [13] implement QC in the standard encryption system after reviewing several security frameworks used to protect health records.

As medical knowledge develops, electronic medical records (EMRs) are frequently employed to increase the efficiency and reliability of healthcare. Since EMRs are reserved distinctly in healthcare and medical institutions, sharing issues arise [14]. Additionally, there is a threat to privacy and security because enormously complex EMRs are susceptible to manipulation and abuse. From a philosophical perspective, the researcher is interested in utilizing cloud storage and remote access to medical treatments in the latest advancements of Healthcare 4.0, which incorporate Internet of Things (IoT) components [15]. The main elements of Healthcare 4.0 were frequent monitoring of medical data, its consolidation, its transfer, its sharing, and its management. There are many challenges in protecting sensitive and private patient information from hackers. To avoid compromise of patient health data by authenticated user elements of electronic healthcare systems, additional security measures are needed once the data is stored, obtained, and transmitted to patients. For the secure storage, transmission, and retrieval of healthcare data across cloud-based service providers, several data encryption techniques have been developed. Table 1 shows an analysis of connected work inside the quantum blockchain space.

4 Advanced Quantum-Integrated Technologies

Quantum computers use neural network-like hardware instead of software. Neurons—the building blocks of neural networks—are qubits. Thus, a qubit-based quantum system can perform neural network functions faster than conventional machine learning methods. To put it another way, with careful application, quantum machines can remove a number of obstacles to the progress of algorithms for machine learning while also potentially improving our quality of life.

QML research seeks to accelerate machine learning algorithms using quantum computing. Quantum-enhanced machine learning (QML) and quantum-assisted machine learning (QAML) are two additional names for QML. The utilization of

Table 1 Quantum blockchain work comparison

Author	Year	QC[a]	BC[b]	QB[c]	HC[d]	Observations
Chen [16]	2022	Yes	Yes	Yes	Yes	The authors created a Blockchain-based Anti-Quantum Attribute-based Authentication for Protected EMRs Sharing
Mahajan [17]	2022	No	Yes	No	Yes	This promotes Healthcare 4.0 innovation by highlighting research gaps, problems, and a future roadmap
Gupta [18]	2022	Yes	Yes	Yes	No	Lattice cryptography was used to resist a quantum attack. A trustworthy blockchain scheme is shown to verify batch data for automobiles
Azzaoui [19]	2022	Yes	Yes	Yes	Yes	This study provides a quantum cloud as a service that is an efficient, scalable, and secure for smart healthcare computations
Mirtskhulava [20]	2021	Yes	Yes	Yes	Yes	Blockchains are viewed as a key 5G solution for mobile IoT security as well EHR exchange. In an effort to increase blockchain security, researchers studied hash-based post-quantum authentication
Iovane [21]	2021	Yes	Yes	Yes	No	We outline a negotiation strategy for blockchain architecture that fixes a transaction's validity and allots a new block Block verification and allocation use expanded likelihood-based negotiation
Fernandez [22]	2020	Yes	Yes	Yes	No	This paper examines post-quantum cryptographic protocols and how they can be used for blockchains and DLT. This article offers a thorough analysis of post-QB blockchain security as well as helpful guidance
Cai [23]	2019	Yes	Yes	Yes	No	The proposed quantum trademark solutions are based on quantum entanglement and employed by either single or multiple signers
Li [24]	2018	Yes	Yes	Yes	No	This study analyzes blockchain networks' quantum-attack vulnerabilities and post-quantum mitigation strategies
Gao [25]	2018	Yes	Yes	Yes	No	Lattice-based signature system was presented. Lattice-based secret delegating creates keys, while preimage survey verifies communications

[a]QC: quantum computing
[b]BC: blockchain
[c]QB: quantum blockchain
[d]HC: healthcare

computer technology for computer vision and other tasks is becoming more common among data analysts and many others as it moves beyond being a work-to-develop and primarily theoretical technology. Today, a few pilot projects at significant IT companies are testing quantum computing in practical settings. The result could result in more practical quantum computing systems as opposed to theoretical ones.

4.1 Quantum Machine Learning

Quantum machine learning (QML), a theoretical field, is just starting to take shape. At the nexus of quantum computing and machine learning, it is situated. The main goal of QML processes is to swiftly learn by combining machine learning with our understanding of quantum computing. The QML employs frameworks from traditional machine learning theory to approach quantum computing. Xanadu provides PennyLane as open-source software designed to simulate quantum machine learning. It combines hardware and quantum simulators with conventional machine learning software. In addition to supporting a growing ecosystem of quantum hardware, PennyLane also supports a wide variety of machine learning frameworks. PennyLane and hardware selections permit commercial clients to start using quantum computing right away, despite the community at large continuing exertions to construct fault-tolerant quantum theory.

4.2 Quantum Drones and Satellites

Implementations for quantum detectors in drones, UAVs, USVs, or UUVs include spectral measurements, surface reflectivity, collision detection in human-made structures (such as passageways or underground homes), and human-based structures (such as delivery-based structures) [26].

Recently, potential results in this area have been observed. Choi et al. described the very first "quantum drone" for secure air-to-ground data transmission [27]. Implementations for quantum detectors in drones, UAVs, USVs, or UUVs include spectral measurements, surface reflectivity, collision detection in human-made structures (such as passageways or underground homes), and human-based structures (such as delivery-based structures) [26]. Recently, potential results in this area have been observed. Choi et al. described the very first "quantum drone" for secure air-to-ground data transmission [27]. Schirber et al. [28] discussed an unmanned aerial vehicle prototype (UAV quantum network prototype that effectively transferred a quantum signal; Schirber et al. [28]). This experiment demonstrates how secure message exchange is possible with quantum communication. Here, a pair of mechanically distinct entangled photons can be shared by two users. The practical choice for sending these photons is optical fibers. Schirber et al. covered the main issues and shortcomings in this area [27]. Quantum satellite-based experiments are covered in [29] [30]. QKD (quantum key distribution) experiments are performed in a short amount of period or at a high rate of speed. The use of quantum satellites for multimedia communications, including audio, video, and text, has also recently been subjected to experimentation and is anticipated to become commonplace in the future.

4.3   Quantum AI

The training of neural networks can be accelerated by combining deep learning and quantum computing. Through the use of this technique, we were able to optimize fundamentally and create a fresh deep learning strategy. On a true, physical quantum computer, we can replicate classical deep learning techniques. The computational complexity of multilayer perceptron topologies grows with the number of neurons. Using highly specialized GPU clusters has the potential to boost performance while significantly reducing training time. In contrast to quantum computers, it will rise.

5   Healthcare Based on Photonic Quantum Computing

Quantum computing will be at the forefront of the subsequent significant change in the medical sector [31]. Many industries can benefit from this technology, but those that have an impact on the health sector especially. Quantum theory and technology enable simulations, drug development, organ molecular deception, and massive data processing. Nanotechnology, AI, and quantum computing enable this. Photonic quantum information has emerged from advances in photon synthesis, control, and sensing. This development includes single-photon switching, correctly fully operational photonic quantum circuits, and state-of-the-art optical metrology that outperforms classical optics. Photonics [32] is known for being a platform for quantum data processing, but it has significant hurdles when it comes to scaling. Large quantum circuits can be built using deterministic methods, but this requires an excessively high number of quantum emitters, whereas nondeterministic methods require enormous resource overheads.

6   Healthcare with Quantum Gates and Circuits

Physicians can add a large amount of cross-functional datasets to their jeopardy factors due to quantum's ability to calculate on scale. Quantum computing will permit the use of additional reference points when picking clinical trial participants, resulting in a more accurate and good fit between the patient and the procedure. Today, providing high-quality care depends on the precision and promptness of diagnosis and treatment. Unparalleled processing power and speed could result from the application of quantum computing. Doctors can make correlations and suggest a diagnosis or treatment using quantum theory. Quantum computing has the ability to totally revolution the way medicine is practiced, including its intelligence. In terms of diagnosing and keeping track of illnesses, quantum computing is not far behind. Cancer patients who receive chemotherapy frequently wait several months to find the outcome of their treatment. However, the development of quantum computing is changing this. To identify the trends that classify them more accurately, physicians can analyze huge amounts of data simultaneously and compare them in comparison along with all conceivable permutations of those data.

It takes time to fully understand how one drug responds to other drugs when taken together. Quantum computing can drastically reduce time because it has the capacity to simulate every possible scenario. The application of quantum computing may enable precise clinical imaging and therapeutic procedures. Because of the precise way quantum imaging devices can produce images, it is now possible to view single molecules. To analyze the effectiveness of a therapy, a doctor can benefit from machine learning and quantum computing. The results of therapy may be easier to interpret over the use of quantum computing; anomalies in the body may be easier to spot with the help of machine learning. In typical magnetic resonance imaging, the radiologist must assess the problem after noticing light and dark areas. However, the ability to differentiate between different types of tissue in quantum imaging methods allows for more thorough and accurate imaging.

7   Healthcare Quantum Algorithms

Healthcare costs are increasing and the volume of data in the industry is increasing. Cognitive computing uses big data, sophisticated machine learning, supercomputers, and cloud services to help clinicians identify diseases early. It is using big data, innovative machine learning, supercomputers and cloud services to help detect cancer early, improve treatment outcomes, and ultimately reduce healthcare costs. By running computations at incredibly fast rates, quantum algorithms enable the solution of computationally challenging problems on conventional computers. This has great potential to obtain important decision-level information from medical images.

Computer scientists develop algorithms that can take advantage of this circumstance to fully utilize the potential of superposition. If you have difficulty comprehending all this, you are not alone. Because quantum theory remains primarily an abstract idea, even the smartest scientists have difficulty comprehending these concepts. Today, a conventional computer can generate the strategy using a limited number of information points in a clinically reasonable amount of time. Quantum-inspired algorithms will permit medical innovations to simultaneously run all variations, use a lot of data, and quickly determine the best course of action.

Quantum-enhanced machine learning techniques are crucial for healthcare in the variety of quantum algorithms. This is because health datasets' regular variability but also unequal distribution may challenge current AI in the near future. Researchers are currently studying methods to enhance the efficiency of computational processes, particularly for big matrices, which are crucial in quantum-based approaches to artificial intelligence and machine learning modeling.

8 Comparing Quantum Computing-Based Healthcare Approaches

Table 2 illustrates current quantum computing approaches for healthcare.

9 Pros and Cons of Quantum Computing-Based Healthcare Approaches

Application of quantum computing to healthcare [42]

1. Increases the safety of the real-time healthcare system.
2. Quicker data processing for authenticated customers.
3. Improves healthcare security for all stakeholders.
4. Both patient care and system design can be improved using smart healthcare scenarios.

Table 2 Comparing quantum computing-based healthcare approaches

Author	Year	A[a]	B[b]	C[c]	D[d]	E[e]	Proposed approach
Ahmad [33]	2022	Y	Y	N	N	Y	A healthcare blockchain framework is detailed
Azzaoui [19]	2022	N	Y	N	N	Y	It is discussed how to secure healthcare data using a quantum cloud ecosystem
Charles [34]	2022	Y	N	N	N	N	Discusses blockchain's current and future applications
Chen [16]	2022	N	Y	N	N	Y	Quantum security is discussed
Grosu [35]	2022	N	Y	N	N	Y	This study demonstrates the current and future need for mobile apps and blockchain in healthcare
Gupta [36]	2022		N	Y	N	N	We talk about quantum computing as well as COVID-19 patient records
Pandey [37]	2022	Y	N	N	N	N	Using non-fungible tokens
Sultana [38]	2022	Y	Y	N	N	Y	The protocols for keeping track of medical records and post-quantum cryptography are discussed
Trenfield [39]	2022	N	Y	N	Y	Y	Digital pharmacy technologies are explored and discussed
Qu [40]	2022	Y	N	N	Y	Y	Medical processing security is emphasized Quantum healthcare data security is also discussed
Zhu [41]	2022	N	Y	N	Y	Y	Quantum-based videoconferencing encryption is discussed

[a]A: quantum blockchain
[b]B: integrated quantum technologies (quantum drones, quantum satellites, quantum AI/ML)
[c]C: photonic quantum computing
[d]D: quantum gates and circuits
[e]E: quantum algorithms

5. Quantum blockchain-based systems are capable of challenge generation, mining security, enhanced access security, and more.

Quantum computing in healthcare care presents several challenges [42]:

1. Environmentally friendly quantum computing may not be possible at scale.
2. It is challenging to use quantum computing in the real world for healthcare, particularly in developing nations.
3. Real-time quantum computing infrastructure is difficult to develop, especially for the medical field.

10 Conclusion and Future Plans

Blockchain technology has proven to be extremely beneficial when dealing with the COVID-19 pandemic and other health systems of similar nature. Modern medical welfare systems involve sensor data that can be utilized to monitor the patient while also preserving their privacy, as well as the secrecy of their health records. With the help of this technology, huge amounts of data can be stored in an orderly and secure manner, so that users can access them whenever they want. This elevates it above all other innovations, placing those latter technologies in an underclass. They can be used whenever and wherever they are needed because they are readily available whenever and wherever they are needed. In a few minutes, patients could be located and tracked using quantum blockchain technology. Data can be concealed inside a quantum blockchain while still remaining safe and accessible. The practice of innovative tools, such as blockchain and quantum computing, could allow hospitals to handle patient data faster without sacrificing their authenticity. There has been a lot of speculation surrounding, respectively, blockchain technology and quantum technology. Whether or not one of these technologies is currently being used in medicine was the objective of this investigation. This field investigates a variety of technologies, such as quantum physics, ratification intelligence blockchain, drones, and machine learning.

References

1. Marella, S. T., & Parisa, H. S. K. (2020). Introduction to quantum computing. In *Quantum computing and communications*. IntechOpen.
2. Kaushik, K., & Dahiya, S. (2022). Scope and challenges of blockchain technology. In *Recent innovations in computing* (pp. 461–473). Springer.
3. Kaushik, K., Dahiya, S., & Sharma, R. (2021). Internet of things advancements in healthcare. In *Internet of things* (pp. 19–32). CRC Press.
4. Singh, M., Dhara, C., Kumar, A., Gill, S. S., & Uhlig, S. (2021). Quantum artificial intelligence for the science of climate change. *arXiv preprint arXiv:2108.10855*.

5. Chugh, N., Kumar, A., & Aggarwal, A. (2016). Security aspects of a rfid-sensor integrated low-powered devices for internet-of-things. In *2016 fourth international conference on parallel, distributed and grid computing (PDGC)* (pp. 759–763). IEEE.

6. Vashisht, S., Gaba, S., Dahiya, S., & Kaushik, K. (2022). Security and privacy issues in IOT systems using blockchain. In *Sustainable and advanced applications of blockchain in smart computational technologies* (pp. 113–127). Chapman and Hall/CRC.

7. Kaushik, K. (2022). Blockchain enabled artificial intelligence for cybersecurity systems. In *Big data analytics and computational intelligence for cybersecurity* (pp. 165–179). Springer.

8. Li, C., Xu, Y., Tang, J., & Liu, W. (2019). Quantum blockchain: A decentralized, encrypted and distributed database based on quantum mechanics. *Journal of Quantum Computing, 1*(2), 49.

9. Kumar, A., de Jesus Pacheco, D. A., Kaushik, K., & Rodrigues, J. J. (2022). Futuristic view of the internet of quantum drones: Review, challenges and research agenda. *Vehicular Communications, 36*, 100487.

10. Kiktenko, E. O., Pozhar, N. O., Anufriev, M. N., Trushechkin, A. S., Yunusov, R. R., Kurochkin, Y. V., Lvovsky, A., & Fedorov, A. K. (2018). Quantum-secured blockchain. *Quantum Science and Technology, 3*(3), 035004.

11. Nandni, C., & Jahnavi, S. (2021). Quantum cryptography and blockchain system: Fast and secured digital communication system. In *Data engineering and intelligent computing* (pp. 453–462). Springer.

12. Singh, K., Kaushik, K., Shahare, V., et al. (2020). Role and impact of wearables in IOT healthcare. In *Proceedings of the third international conference on computational intelligence and informatics* (pp. 735–742). Springer.

13. Bhavin, M., Tanwar, S., Sharma, N., Tyagi, S., & Kumar, N. (2021). Blockchain and quantum blind signature-based hybrid scheme for healthcare 5.0 applications. *Journal of Information Security and Applications, 56*, 102673.

14. Jhang, J.-H., & Lian, F.-L. (2020). An autonomous parking system of optimally integrating bidirectional rapidly-exploring random trees* and parking-oriented model predictive control. *IEEE Access, 8*, 163502–163523.

15. Kumar, A., Ottaviani, C., Gill, S. S., & Buyya, R. (2022). Securing the future internet of things with post-quantum cryptography. *Security and Privacy, 5*(2), 200.

16. Chen, X., Xu, S., Qin, T., Cui, Y., Gao, S., & Kong, W. (2022). Aq–abs: Anti-quantum attribute-based signature for EMRS sharing with blockchain. In *2022 IEEE wireless communications and networking conference (WCNC)* (pp. 1176–1181). IEEE.

17. Mahajan, H. B., Rashid, A. S., Junnarkar, A. A., Uke, N., Deshpande, S. D., Futane, P. R., Alkhayyat, A., & Alhayani, B. (2022). Integration of healthcare 4.0 and blockchain into secure cloud-based electronic health records systems. *Applied Nanoscience, 126*, 1–14.

18. Gupta, D. S., Karati, A., Saad, W., & da Costa, D. B. (2022). Quantum-defended blockchain-assisted data authentication protocol for internet of vehicles. *IEEE Transactions on Vehicular Technology, 71*(3), 3255–3266.

19. Azzaoui, A. E., Sharma, P. K., & Park, J. H. (2022). Blockchain-based delegated quantum cloud architecture for medical big data security. *Journal of Network and Computer Applications, 198*, 103304.

20. Mirtskhulava, L., Iavich, M., Razmadze, M., & Gulua, N. (2021). Securing medical data in 5g and 6g via multichain blockchain technology using post-quantum signatures. In *2021 IEEE international conference on information and telecommunication technologies and radio electronics (UkrMiCo)* (pp. 72–75). IEEE.

21. Iovane, G. (2021). Murequa chain: Multiscale relativistic quantum blockchain. *IEEE Access, 9*, 39827–39838.

22. Fernandez-Carames, T. M., & Fraga-Lamas, P. (2020). Towards post-quantum blockchain: A review on blockchain cryptography resistant to quantum computing attacks. *IEEE Access, 8*, 21091–21116.

23. Cai, Z., Qu, J., Liu, P., & Yu, J. (2019). A blockchain smart contract based on light-weighted quantum blind signature. *IEEE Access, 7*, 138657–138668.

24. Li, C.-Y., Chen, X.-B., Chen, Y.-L., Hou, Y.-Y., & Li, J. (2018). A new lattice-based signature scheme in post-quantum blockchain network. *IEEE Access, 7*, 2026–2033.
25. Gao, Y.-L., Chen, X.-B., Chen, Y.-L., Sun, Y., Niu, X.-X., & Yang, Y.-X. (2018). A secure crypto currency scheme based on post-quantum blockchain. *IEEE Access, 6*, 27205–27213.
26. Jordan, S. P., & Liu, Y.-K. (2018). Quantum cryptanalysis: Shor, Grover, and beyond. *IEEE Security & Privacy, 16*(5), 14–21.
27. Choi, C. (2019). World's first "quantum drone" for impenetrable air-to-ground data links takes off. *IEEE Spectrum.*
28. Schirber, M. (2021). Quantum drones take flight. *Physics, 14*, 7.
29. Huang, D., Zhao, Y., Yang, T., Rahman, S., Yu, X., He, X., & Zhang, J. (2020). Quantum key distribution over double-layer quantum satellite networks. *IEEE Access, 8*, 16087–16098.
30. Raska, M. (2016). China's quantum satellite experiments: Strategic and military implications. *Nanyang Technological University. RSIS, 223.*
31. Takeuchi, S. (2016). Photonic quantum information: Science and technology. *Proceedings of the Japan Academy, Series B, 92*(1), 29–43.
32. Bartlett, B., Dutt, A., & Fan, S. (2021). Deterministic photonic quantum computation in a synthetic time dimension. *Optica, 8*(12), 1515–1523.
33. Ahmad Naqishbandi, T., Syed Mohammed, E., Venkatesan, S., Sonya, A., Cengiz, K., & Banday, Y. (2022). Secure blockchain-based mental healthcare framework: A paradigm shift from traditional to advanced analytics. In *Quantum and blockchain for modern computing systems: Vision and advancements* (pp. 341–364). Springer.
34. Charles, W. M. (2022). The future of blockchain. In *Blockchain in life sciences* (pp. 315–336). Springer.
35. Grosu, G.-M., Nistor, S.-E., & Simion, E. (2022). A note on blockchain authentication methods for mobile devices in healthcare. *IACR Cryptology ePrint Archive, 2022*, 159.
36. Gupta, S., Modgil, S., Bhatt, P. C., Jabbour, C. J. C., & Kamble, S. (2022). Quantum computing led innovation for achieving a more sustainable covid-19 healthcare industry. *Technovation, 120*, 102544.
37. Pandey, S. S., Dash, T., Panigrahi, P. K., & Farouk, A. (2022). Efficient quantum non-fungible tokens for blockchain. *arXiv preprint arXiv:2209.02449.*
38. Sultana, T., Mazumder, R., & Su, C. (2022). Post-quantum signature scheme to secure medical data. In *Rhythms in healthcare* (pp. 129–146). Springer.
39. Trenfield, S. J., Awad, A., McCoubrey, L. E., Elbadawi, M., Goyanes, A., Gaisford, S., & Basit, A. W. (2022). Advancing pharmacy and healthcare with virtual digital technologies. *Advanced Drug Delivery Reviews, 182*, 114098.
40. Qu, Z., Zhang, Z., & Zheng, M. (2022). A quantum blockchain-enabled framework for secure private electronic medical records in internet of medical things. *Information Sciences, 612*, 942–958.
41. Zhu, D., Zheng, J., Zhou, H., Wu, J., Li, N., & Song, L. (2022). A hybrid encryption scheme for quantum secure video conferencing combined with blockchain. *Mathematics, 10*(17), 3037.
42. Kaushik, K., & Kumar, A. (2022). Demystifying quantum blockchain for healthcare. *Security and Privacy, 6*, 284.

Sustainable Solutions for Serverless Edge, Fog, and Cloud Computing Using Quantum and Blockchain Technology

Sarthika Dutt, Vansh, Garima Pahwa, Aishvi Guleria, Kamya Varshney, and Deeksha Joshi

Abstract With the advent of emerging technologies, quantum and blockchain security and privacy concerns on various platforms have been solved effectively. Additionally, the advancements in these research areas are disrupting every industry. These technologies and their integration with other several technologies have encouraged experts of different domains to incorporate technological solutions for several requirements. Quantum and blockchain together can provide sustainable and resilient applications that can be utilized effectively in various domains. This is the purpose for growing investment on quantum and blockchain based computing by the government and private firms. The present study analyzed the quantum science and blockchain technology scope when combined with different computing paradigms. Additionally, we explored the continuous advancement of new blockchain technologies and quantum computing algorithms that provide protected quantum solutions, quantum proofs, and anti-quantum schemes was determines any scheme for quantum attacks and generate a secure quantum computing model. The implications of quantum resilient blockchain technologies can lead futuristic events to another level. Therefore, this chapter discusses the practical examples to emphasize effects of blockchain and quantum computing paradigms. Potential challenges in quantum and blockchain sustainability have been detailed. This work also explores the potential of quantum and blockchain to foster deep sustainability transformations.

Keywords Quantum computing · Blockchain · Security · Sustainability · Cloud computing · Edge computing · Fog computing

S. Dutt (✉) · Vansh
COER University, Roorkee, India

G. Pahwa · A. Guleria · K. Varshney
Indira Gandhi Delhi Technical University for Women, New Delhi, India

D. Joshi
Graphic Era Hill University, Dehradun, India

© The Author(s), under exclusive license to Springer Nature Switzerland AG 2024
S. Pulipeti et al. (eds.), *Quantum and Blockchain-based Next Generation Sustainable Computing*, Contributions to Environmental Sciences & Innovative Business Technology, https://doi.org/10.1007/978-3-031-58068-0_10

"}

1 Introduction

Quantum computing is an emerging technology that provides solution to problems that are too complex for classical computer. The best features of this technology are that it has more prospective to solve difficulties than conventional methods. Quantum computing uses their own bits to store and update the information which is also known as qubits. Whereas classic computing uses the binary bits 0 and 1, on the other hand, quantum computing uses atoms, photons, electrons, and ions as its bits with its critical information. It helps to super put and gives further combinations. They can run using memory effectiveness; therefore, they're more important. Only the quantum computing model disobeys the Church-Turing thesis. This is the main reason that quantum computers can accomplish any task quicker more efficiently than classical computers [1]. Whereas the characteristics of blockchain provide transparency, security, immutability, and rigidity. It has been applied in numerous areas similar as cryptocurrency, equity backing, and commercial governance. In experimental stage, the blockchain technology has to face problems which need to be answered including data processing capacity, information intimately, and nonstop difficulties. This blockchain technology knowledge works effectively to give security and translucency. The public blockchain could be used as a place for enterprises to freely hide information. For long duration, the features of blockchain technology helped to abate miscalculations in exposure and earnings operation and better quality of counting information and alleviate information asymmetry [2]. The blockchain technology could also be utilized in the artificial intelligence and machine learning areas. The organizational quality could also be enhanced by incorporating blockchain technology [3–9].

The advancement in serverless technology has its own effect on the present industrial revolution 4.0. Cloud, edge, and fog computing are some of the serverless computing paradigms enormously utilized by the industries, organizations, and individuals. Cloud computing stores the data and programs through the Internet rather than hard drive. It is the use of software to deliver services through network connectivity such as cloud computing provider Google's Gmail [10]. The edge computing here indicates nonfictional geographic distribution. Edge computing is accomplished adjoining the source of data, instead of counting on a dozen data centers to complete tasks. Fog computing shift is a centralized, distant pall, toward original and end-user bias. Fog computing promotes scalability and handles data [11, 12]. The prime thing required here is to find an equilibrium between operations of coffers and waitpersons without losing performance and services delivered [13]. The big data provides connectivity to the structured and unstructured data, at an unknown rate. The relationship and correlational analysis are all performed based on data that's why it reveals the unseen connections within the vast ocean of data. The development in quantum and blockchain has put the present technology in the condition where data is in continuous threat to get exploited. Most of the technologies, devices, and data would be in vulnerable conditions with the development in quantum computers. Therefore, this study presents the data sustainability required with these emerging technologies. The sustainable solutions for serverless computing technologies utilizing quantum and blockchain technology are explored in the present study. The paper organization is as follows: the next section covers the quantum and blockchain overview, followed by its effect on computing paradigms; next section explores sustainable computing paradigms; and finally, the future direction and conclusion of this work are discussed.

2   Quantum and Blockchain Overview

A quantum computer consists of quantum bits (qubits) as basic units to store information. These qubits adhere to the superposition principle and can be in binary zero and one states (classical bits), also can attain any state between zero and one.

$$0i + ?1i, \tag{1}$$

where $\alpha, \beta \in C$,

$$\text{Subjected to} 2 + 2 = 1 \tag{2}$$

A subset of distributed ledger technology (DLT) is blockchain. It's a distributed database that keeps track of assets and network-wide transactions; it's a decentralized database. The identical copies of records are shared by every node in the blockchain system, creating disruption and modification practically impossible. In a network, every node or system is connected peer to peer (P2P). This distribution is at the level of a theoretical network. A blockchain replicates information over a vast network of distributed computers as opposed to storing it in a single location. This is accomplished by employing a network of specialized computers to resolve a cryptographic algorithm and authenticate transactions between two parties. The answer is then incorporated in a block that is further added to the record. Blocks are universally accessible and publicly accessible recordings of blockchain transactions that are not hosted by a single person. Figure 1 depicts the structure of qubits in quantum computing paradigms.

2.1   *Blockchain Vulnerabilities to Quantum Attacks*

Any system that depends on mathematical methods for security is very much at risk from quantum technologies. Blockchain is particularly vulnerable since one-way functions serve as its primary line of defense and a user's digital signature is their only form of defense. The most immediate threat is consequently the cracking of digital signatures. Shor's technique can be used by a criminal using a quantum computer to forge any digital signature, pose as any user, and take their digital assets.

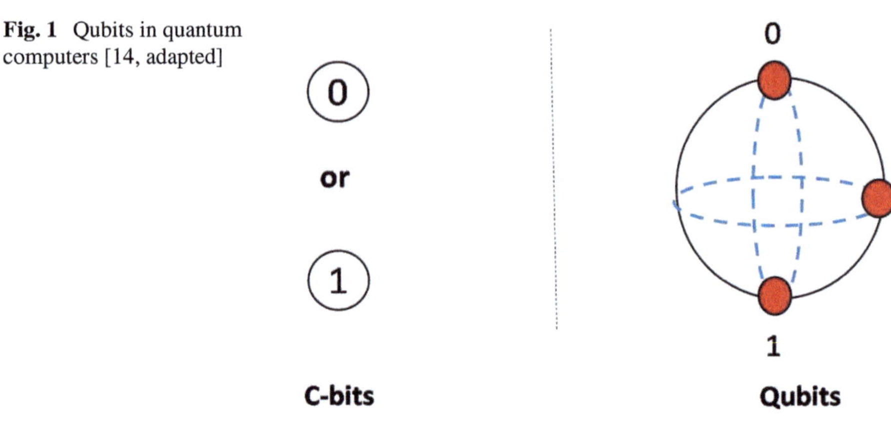

Fig. 1 Qubits in quantum computers [14, adapted]

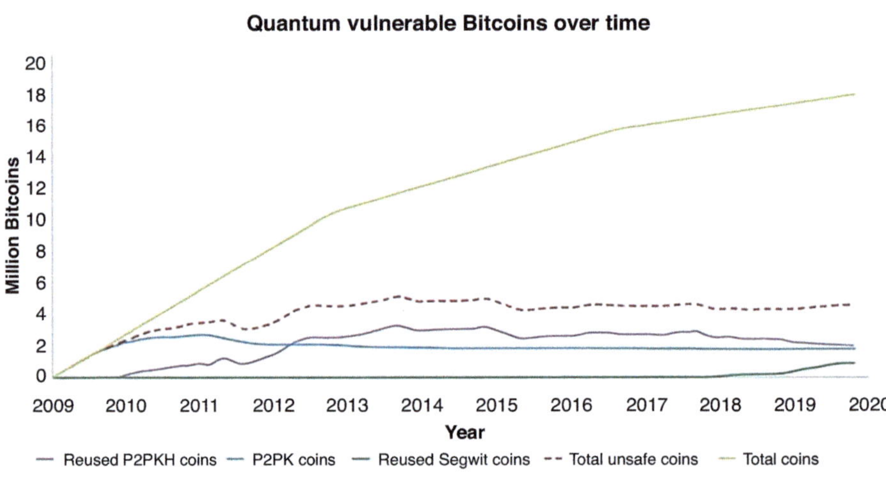

Fig. 2 Graph showing blockchain vulnerability to quantum computing (https://www2.deloitte.com/nl/nl/pages/innovatie/artikelen/quantum-computers-and-the-bitcoin-blockchain.html)

One-fourth of the Bitcoins in leverage are susceptible to a quantum attack [14]. Quantum computers may one day be able to effectively calculate one-way functions like the Internet and the blockchains used to secure financial transactions. The extensively utilized one-way encryption will be shortly rendered obsolete. By 2035, quantum computer would be able to crack the critical RSA2048 cryptographic system. Figure 2 represents the blockchain vulnerability to quantum computing over the years. All the public transactions are vulnerable as anyone can access public key from p2pk address. A hash of the public key makes up the recipient's address but does not reveal the public key. Pay to Public Key Hash (p2pkh), the first and most well-known implementation of this, prevents retrieval of the public key from the address. This means that the public key is unknown, and the private key cannot be obtained using a quantum computer if money has never been transmitted from a p2pkh address. The public key is disclosed if money is ever moved from a specific p2pkh address. This address might then be used to obtain the private key using a quantum cutter running Shor's algorithm. As a result, a rival with a quantum computer would be able to use the address's coins.

According to existing scientific assessments, a quantum computer will take around 8 h to crack an RSA key, whereas a Bitcoin signature might be broken down for approximately 30 min. Therefore, Bitcoin ought to be resistant to quantum attacks (as long as address is not reused).

2.2   Quantum Based Blockchain Security

Quantum technologies not only exploit blockchain security but also can provide opportunities for security as well.

1. *Quantum-safe encryption.* No user can impersonate another in quantum communications since they are automatically authenticated. Such systems communicate and encrypt information using the states of individual light particles (photons). Quantum states can neither be replicated nor measured without causing an alteration, according to basic physics. Any listener will be discovered right away. Classical digital signatures can be replaced by quantum cryptography, which can also be used to encrypt all peer-to-peer conversations in the blockchain network.
2. Owing to the nature of quantum computing, the SVP is challenging for quantum computers to solve. A quantum computer can only employ the superposition principle when all of the qubit states are perfectly aligned. However, when the states are not, it must turn to more traditional methods of computation. Therefore, it is extremely improbable that a quantum computer will be able to solve the SVP. Lattice-based encryption is hence resistant to quantum computers.

2.3   Quantum-Resistant Blockchain

A digital quantum-resistant algorithm is implemented in quantum-resistant computer blockchain system. However, it lacks actual experience and only engages in theoretical research. The signed public key consumes a significant amount of block space. The issue of the lengthy public key is still not fully resolved as of right now.

2.4   *CPS 4.0-Based Blockchain with Quantum Resistance*

Some scientists are developing digital signature algorithms. A digital signature algorithm based on bonsai trees was proposed by Li et al. [15]. Its security can be ensured by this algorithm. It is, however, ineffective. Additionally, its viability must be confirmed. A blockchain-compatible double signature approach was put forth by Gao et al. [16]. However, this scheme's security is only guaranteed by the unconvincing SIS assumption.

Yin et al. [17] extend a lattice space to many lattices on the basis of the bonsai tree. The signature becomes more complex due to this method. Additionally, these systems yield a huge number of signatures [6]. Public blockchain works best when a network that cannot be trusted at all has to be secure, but it is slow and expensive. For instance, it actually costs money to set up a powerful mining node, and processing the mining operations uses a lot of energy. Furthermore, while real-time data exchanges are necessary in many industries, public blockchain is incompatible with certain use cases because it can only validate a small number of transactions per second. Since using a token for incentives or rewards is not required in this case, mining setup costs can be reduced, making the system more suitable for real-time, mission-critical systems like Industry 4.0 applications. Permissioned blockchain (BC) has a greater transaction processing rate and is cheaper and faster [14].

3   Quantum and Blockchain Effect on Computing Paradigms

Computing application plays an important role and is continuously evolving with technological advancements. It provides the users the capability to transfer data from remote location to other remote locations with limited resources. However, the developments in quantum computers and blockchain technology are somewhere supporting its growth but also create vulnerabilities for the same. The impact of these technological advancement is discussed under this section.

3.1   *Impact of Serverless Computing in Delivering Cloud Services Through IoT Based Applications*

Serverless solutions have new challenges as a result of the Internet of Things, including managing specific situations, little overhead, quick answers, and strict resource usage. Serverless computing is the development and deployment of applications as a collection of stateless functions (FaaS). Serverless computing, which was primarily designed for the cloud, has also found a place on the Internet of Things by bringing services closer to the devices in order to decrease latency and eliminate unnecessary energy and resource use. It's intriguing how well some systems can function across cloud, fog, and edge layers. Powerful cloud-based algorithms can be used to process the massive amount of data produced by IoT sensors as long as there is an Internet connection. The process of managing cloud services is by no means simple. A user has a number of difficulties when administering a cloud environment. For instance, the cloud user must guarantee service availability so that if one computer fails, the entire service is not affected. Additionally, in order to secure the services from disasters, he or she must think about geographically dispersing copies of the services. Load-balancing is an additional difficulty. To effectively utilize all resources in this situation. The serverless cloud computing model was developed in response to these difficulties. In Fig 2 serverless cloud computing is depicted as providing backend as a service (BaaS) and function as a service (FaaS). The FaaS, however, enables programmers to deploy and run their code on computer platforms. The BaaS's services, including those for a database, messaging, user authentications, and other things, are necessary for the FaaS to function. It is also referred to as "event-driven functions" and is regarded as the most popular serverless architecture. Amazon Lambda offered the serverless cloud concept for the first time in 2014, and cloud providers like Google and Microsoft followed suit in 2016.

Another aspect of serverless computing is paying for resources as they are utilized. The developer fees in this cloud computing paradigm are based on actual resource usage. A serverless provider only gets paid when an application starts using resources; therefore, deploying an application won't cost the developer anything if it isn't being used. Any new technology will have a lot of operational and technical issues and difficulties in the beginning. Recent developments in serverless cloud computing have brought forth a variety of problems. Serverless cloud computing lacks tools for managing and keeping an eye on serverless applications. It might also involve security concerns. Vendor lock-in is a problem for the providers of serverless services as well. Nevertheless, despite the fact that it has not been extensively researched in academic studies, serverless cloud computing has attracted favorable attention in the industry.

3.2 Edge Computing to Mitigate Latency in Activity Oriented Serverless Computing

With the rise in intelligent IoT edge applications, a significant amount of computing resources is being used to run these IoT edge apps. The resource required for these applications cannot be met by typical methods of shifting work to the cloud due to the low bandwidth and significant communication latency between IoT devices and cloud servers. Edge computing provides services to these applications within network facilities, computational power, and storage capacity. This improves network transmission rate as well as network congestion, privacy protection, and security. When an edge server is overloaded, resource contention and congestion may happen, increasing latency or possibly leading to task failure. To ensure the effective use of resources and the system performance of edge servers, various job demands must be fulfilled in IoT devices. Making an effective task offloading strategy in edge computing, then, to lower task processing latency and thereby improve performance, is a significant challenge. It has been suggested to integrate serverless edge computing, often referred to as function as a service (FaaS), with edge computing in order to boost edge computing performance and make it simpler to create and implement event-driven IoT applications. A virtual machine remains online after completing a process and waits for a new task to arrive. This is known as a stateful execution model. A stateless execution approach is used in serverless computing, and the container is destroyed after the operation has been completed to prevent resource waste. Edge FaaS (EFaaS) is made available by the two primary cloud platform providers for function execution at edge devices. To assure long-term performance optimization, techniques have recently been researched. Existing DRL-based work offloading strategies face obstacles, nevertheless, like low sample variety and large exploration costs. The serverless edge computing architecture was developed to improve the efficiency of edge computing and make it easier to design and deploy event-driven IoT applications. It blends edge computing with serverless computing, sometimes referred to as function as a service (FaaS), and is used for edge AI inference systems.

4 Sustainable Blockchain and Quantum Computing Paradigms

Seventeen Sustainable Development Goals (SDGs) were set by the United Nations (UN) member states in 2015, and they must be accomplished by 2030 (http://www.cepal.org/en/node/37174). In order to establish a roadmap for economic equality in a sustainable society, the SDGs constitute a pressing global call to action. The European Green Deal, a blueprint for carrying out the UN 2030 agenda with a dedication to a growth strategy that would turn environmental difficulties into possibilities across all policy areas, was introduced at the European level in December 2019 [18]. Since it can deliver transparency, accountability, reliability, and information security, higher operating block chain technology serves as one of the critical empowering technologies that can assist to create safe and sustainable solutions to achieve these SDGs.

4.1 Quantum and Blockchain Based Cloud Computing

By outsourcing their necessary services, organizations using cloud computing can reduce overall costs. Because of exporting, it poses new challenges in terms of reliability, integrity, and secrecy for data protection. One of the most popular expansions today that possibly address security concerns with cloud computing is inclusion of blockchain technology. For commuters, reliance on a central system for data management and decision-making generates a severe dilemma. For illustration, if a central server crashes, the entire system may suffer, and critical data retained on the server may be compromised. The primary server is also susceptible to hacker stabbings. The blockchain can deliver a solution to this problem since it has a decentralized design that, maintaining many copies of the same data on different computer nodes, prevents the entire system from failing if one server collapses. Furthermore, data loss is not a problem because there are backups.

Use Case 1: Smart Manufacturing
A developing area where BCoT can contribute is smart manufacturing. Smart manufacturing involves the use of automated devices that can carry out particular jobs far more intelligently than they can now. This industry is founded on service-oriented manufacturing, technology backed by IoT, and cloud manufacturing. But given the current situation, centralized industrial networks and third party-based authority are issues that smart manufacturing must deal with. Reduced adaptability, effectiveness, and security are the results of centralized production designs. The implementation of BCoT, is one potential remedy for these problems in the decentralized approach, which can increase security [19].

Use Case 2: Smart Home Automation
Smart home automation is one of the key industry sectors where blockchain-enabled BCoT can be applied. By automating the home's appliances, a house becomes a smart house that provides comfort for the occupants. A network of Internet of Things (IoT) devices is a collection of gadgets like sensors and detectors; these gadgets collect information from their surroundings, store it on the network, evaluate it, and then perform a certain task based on the pertinent data. For instance, a temperature sensor in a fire alarm system assesses whether there is a fire within the home. If a fire is discovered after this is detected, processed, and analyzed, a message is transmitted to the home's owner, and house's installed water sprinklers or fire alarms are automatically triggered [20]. Automated blockchain could be utilized in homes to enhance the security by eliminating the risk of losing data or concern for privacy of data. Without the requirement for a third party, a secure information integrity architecture using blockchain can ensure the security and stability of the whole network.

4.2   Quantum Science and Blockchain Technology for Serverless Edge Computing/Fog Computing

Security and computational performance became a concern for serverless edge computing as a result of enormous data processing. Blockchain and quantum are new-age solutions that have the potential to reduce the security threats and increase the speed. The increase in compute speed can chalk up to blockchain as it processes data in blocks where quantum computing processes huge data for load-balancing as well as dynamic provisioning.

Use Case 1: Energy Management
Running the necessary massive facilities demands a lot of energy. Using quantum computers and the blockchain concept, the proposed model can be used to increase the service's security, speed of processing, and dependability. Computing paradigms of this technological era with innovative memory architecture may help in reducing the energy consumption levels suitable for energy-saving techniques and judicious use of renewable resources. While processing data from Internet of Things applications, software-defined network (SDN)-based fog and edge computing can reduce energy consumption and boost network usage [21].

Use Case 2: IoT and 5G/6G Technologies
A rapid data processing method with minimal latency and reaction time is required for IoT applications which find their use in smart city, weather forecasting, health, and agriculture sectors. Serverless edge computing using quantum computers can offer powerful computation for quick data processing at edge service. Owing to this, 5G (fifth generation) and 6G (sixth generation) telecommunication systems were devised, and they are able to deliver large capacity data transfer rates with infinitely small delay with low consumption of power. They have enabled quick gatherings and distribution of data [22].

Use Case 3: AI
Utilizing cutting-edge optimization and AI-powered approaches is necessary to enable dynamic IoT environments and boost computing systems' efficiency [14, 23, 24]. Additionally, by employing extensible machine learning techniques and strategic planning policies, AI can help in improving quality of service (QoS) [25].

4.3   Quantum and Blockchain Based Osmotic Computing

The main objective of this new generation of computing is to maintain resource allocation and load balance among servers without impairing service delivery or performance. Osmotic computing which mixes IoT, cloud, fog, and edge computing by dynamically managing IT services has proven to be a popular concept. Osmotic computing has established itself in a variety of contexts, including linked autos,

Industry 4.0, smart cities, and smart healthcare. In order to manage resources and resolve data difficulties in the osmotic computing, data science, IoT pivots on allocation of services of system using the fundamental chemistry principles of the osmosis phenomena. Additionally, it establishes a foundation for a dynamic ecosystem in which vehicles, wayfinding devices, and physical infrastructure constantly exchange information on city traffic.

Use Case 1: Blockchain Enabled Osmotic Manager
Existing technology of osmotic computing can be strengthened by involvement of blockchain. It can establish reliability, build trust on the system, and increase transparency of the model. Blockchain minimized the third-party involvement, and taking advantage of this feature of blockchain, it can be integrated with OC such that when multiple vendors run on the same Multi-Cloud of Things, a transparent environment is maintained. Here the need for executing mutually agreed policies is highlighted. So smart contract which is another boon of blockchain comes into play. Similarly, trust can be enforced through consensus algorithms [26].

4.4   Quantum and Blockchain Big Data Computing

In the shortest possible time, big data organizes and extracts valuable information from rapidly expanding, massive volumes of data in various configurations and combinations, regularly updated and collected from multiple independent sources. Big data can be defined by the five Vs: variety, volume, veracity, velocity, and value. This technology of blockchain is capable to provide highly protected transaction of data in bulk due to its decentralized and immutable nature. In an effort concerning data privacy that still exists in regard to already established methods of transmission of data, it makes reliable data transmission between data sources and data analytics possible [27]. In order to carry out the transmission of data, blockchain can guarantee big data training and prevention in the theft of data.

Use Case 1: Smart Transportation
The fusion of blockchain technology with a variety of other technologies, such as mobility as a service, has altered the conventional approach to transportation (MaaS), artificial intelligence, deep learning, and IoT [37]. The automobile industry has also utilized blockchain to create highly efficient and intelligent systems of transportation and make services like ride-sharing, automatic insurance, and smart charging readily available. Another ability of this technology is to address a number of issues related to auto insurance. This enables insurers to be able to trace their claims easily by examining ledger [28].

Use Case 2: Smart Healthcare
The healthcare sector needs validation, and blockchain delivers that function in a way that all sides can trust. While no one entity is in responsibility of maintaining the data, everyone involved is in a position to ensure its security and integrity. The

healthcare sector may avoid two of its most serious big data concerns at once if no one can update the record without the consent of all stakeholders, and no one can access the medical record without the participants' cooperative authorization.

And the notion that there is, in fact, a widely accessible single fact that is instantly disseminated among stakeholders and doesn't rely on manual data verification subject to human error will eventually help patient care.

5 Challenges in Blockchain and Quantum Computing Paradigms

There are some downsides of amount computing, and inventors should do cautiously when experimenting with operations that involve sensitive data. Also, a machine may not be incontinently available when an inventor wants to submit a job through amount services on the public pall.

Challenges within blockchain and pall computing

1. Technical complexity. Integrating blockchain technology with pall computing requires complex fine processes to transfer the data, process it, and secure the network. This requires important computers, which consume a lot of electricity, and professed mortal coffers. As a result, all these factors produce a bottleneck for large-scale handover due to the high costs involved [35, 36, 40].
2. Lack of skilled workforce. This is still an arising and incipient technology, so a rightly trained labor force cannot execute these processes. As a result, demand for such a professed pool is high and force is low [29].
3. Horizonless incremental model. There were 100 million Qtum released at the time of the mainnet's introduction. Only a subset of these commemoratives has been switched to mainnet commemoratives since the mailjet's introduction. The mainnet also brought with it Qtum mining and the new Qtum that was produced from it in the form of block pricing [30].
4. Scarcity with EVM. Since Solidity is still under development, various new issues are consistently found. Big data quickly organizes and extracts valuable information from rapidly growing, massive datasets in various configurations and combinations, which are regularly updated and collected from multiple independent sources[31].
5. Staking with lower loss. While Ethereum is proposing a minimum of 32 additional participants to share in the staking process, Qtum has no restrictions on a minimum quantity of Qtum for staking [32–34].

6   Future Research Directions

Organization and industries have realized the need of digital business structure based on decentralized structure. Serverless technology supports and shifts the computing paradigm from configuration and management to run operations. The benefits of serverless technology include offering coffers in a pay-as-you-go manner, which reduces costs effectively. But there are numerous challenges that are required to overcome before this can be regarded as a comfortable platform for scalable analytics. The blockchain various features can be veritably profitable in the field of functionality and performance improvement. Blockchain being a decentralized system provides help in icing an armature in which computers can work simultaneously on a particular task and on the processing of data that eventually reduces operation duration and enhances processing speed, uploading, and reacquiring data. There's an aspect of better security in pall computing. Pall computing generally deals with a veritably large quantum of data, so there's always a threat factor of data instability as pall computing works on centralization, and in case the whole centralized garcon gets addressed, the whole system will collapse, and in case there's no data coagulate, the data is lost. So, kinds of issues can be provisioned to by using pall computing and blockchain together. Decentralization company's reliance on a centralized garcon for managing data and making opinions is a big problem in the request. If there's some problem in the central system, it can beget problems in the whole system making it delicate to prize data. Also, hackers can attack and lose the central garcon which fluently gives them access to all the data related to a central garçon. Still, in the case of blockchain and pall computing technology, we use the concept of decentralization in which data is stored in colorful computer bumps. This means if one garcon fails, it won't fail the whole system. Also, blockchain-grounded distributed pall computing mitigates the threat of data loss and ensures data security. Hence, pall-grounded blockchain services offer briskly disaster operation. Storing any company's data in the cloud is a major concern because it often includes sensitive information such as personal details, video footage, voice recordings, and much more. Any leak of similar data can be detrimental to the character of the company and can be dangerous. Still, numerous blockchain pall storehouse companies offer blockchain-grounded pall storehouse results that can greatly ameliorate security [38, 39]. The operation of decentralized pall storehouse blockchain results divides the stoner's data into several translated parts and links them through a mincing function. Each member is stored in a decentralized position.

7   Conclusion

The advancement in quantum and blockchain technology has the advantage of both technologies together to provide secure, transparent, and effective applications. However, both the technologies are at the initial phase of their development but could be profitably incorporated into the commercial sector later. The quantum technology could exploit present security concerns in no time. As a result, the social communities get affected in the process to overcome these limitations. Therefore, quantum and blockchain based serverless computing resources are still an area required to be explored with significant research studies. This study has focused on the serverless edge, fog, and cloud computing use cases and applications to explain the expected outcomes as the commercial applications. However, there are challenges and limitation present in the commercial usage of these technologies together.

References

1. Feynman, R. P. (1982). Simulating physics with computers. *International Journal of Theoretical Physics, 21*(6/7), 133.
2. Antonopoulos, A. M. (2017). *Mastering bitcoin: Programming the open blockchain.* O'Reilly Media.
3. Golosova, J., & Romnovs, A. (2018). Overview of the blockchain technology cases. In *Proceedings of the 59th international scientific conference on information technology and management science of Riga Technical University (ITMS), October 10–12, 2018, Riga, Latvia* (pp. 1–6). IEEE. ISBN 978-1-7281-0098-2.
4. Pilkington, M. (2016). *Blockchain technology: Principles and applications.* Edward Elgar Publishing.
5. Franko, C. (2017). *Borderless: A governance platform and charity for a global society.* Research Gate.
6. DHL Trend Research. (2018). Blockchain in logistics.
7. Ascribe. Lock in attribution, securely share and trace where your digital work spreads. [online]. Available from: https://www.ascribe.io/
8. Everledger. Pioneers of digital provenance. [online]. Available from: https://www.everledger.io/about-us/about
9. MedRec. What is Medrec? [online]. Available from: https://medrec.media.mit.edu/
10. Mell P, Grance T, et al. (2011). *The NIST definition of cloud computing* [Internet]. [cited 18 Sep 2017]. National Institute of Standards and Technology. Report No.: Special Publication 800–145.
11. Satyanarayanan, M. (2017). The emergence of edge computing. *Computer, 50*(1), 30–39. https://doi.org/10.1109/MC.2017
12. Ranjan, R., Rana, O., Nepal, S., Yousif, M., James, P., Wen, Z., Barr, S., Watson, P., Jayaraman, P., Georgakopoulos, D., Villari, M., Fazio, M., Garg, S., Buyya, R., Wang, L., Zomaya, A. Y., & Dustdar, S. (2018). The next grand challenges: Integrating the Internet of Things and data science. *IEEE Cloud Computing, 5*(3), 12–26. https://doi.org/10.1109/MCC.2018.032591612
13. Gamal, M., Rizk, R., Mahdi, H., & Elnaghi, B. E. (2019). Osmotic bio-inspired load balancing algorithm in cloud computing. *IEEE Access, 7*(2019), 42735–42744. https://doi.org/10.1109/ACCESS.2019.2907615
14. Kumar, A., Ahuja, N. J., Thapliyal, M., Dutt, S., Kumar, T., Augusto, D., Konstantinou, C., & Choo, K.-K. R. (2023). Blockchain for unmanned underwater drones: Research issues, challenges, trends and future directions. *Journal of Network and Computer Applications, 215,* 103649–103649. https://doi.org/10.1016/j.jnca.2023.103649
15. Fang, W., Chen, W., Zhang, W., et al. (2020). Digital signature scheme for information non-repudiation in blockchain: A state of the art review. *Journal on Wireless Communications and Networking, 2020,* 56. https://doi.org/10.1186/s13638-020-01665-w

16. Gao, Y., Chen, X., Sun, Y., et al. (2018). A secure cryptocurrency scheme based on post-quantum blockchain. *IEEE Access, 6*, 27205.

17. Yin, W., Wen, Q., Li, W., Zhang, H., & Jin, Z. (2018). An anti-quantum transaction authentication approach in blockchain. *IEEE Access, 6*, 5393–5401. https://doi.org/10.1109/access.2017.2788411

18. Punathumkandi, S., Sundaram, V. M., & Panneer, P. (2021). Interoperable permissioned-blockchain with sustainable performance. *Sustainability, 13*, 11132. https://doi.org/10.3390/su132011132

19. Li, Z., Barenji, A. V., & Huang, G. Q. (2018). Toward a blockchain cloud manufacturing system as a peer to peer distributed network platform. *Robotics and Computer-Integrated Manufacturing, 54*, 133–144. https://doi.org/10.1016/j.rcim.2018.05.011

20. Khan, F., Xu, Z., Sun, J., Khan, F. M., Ahmed, A., & Zhao, Y. (2022). Recent advances in sensors for fire detection. *Sensors (Basel), 22*(9), 3310. https://doi.org/10.3390/s22093310

21. Himeur, Y., Elnour, M., Fadli, F., Meskin, N., Petri, I., Rezgui, Y., Bensaali, F., & Amira, A. (2022). AI-big data analytics for building automation and management systems: A survey, actual challenges and future perspectives. *Artificial Intelligence Review, 15*, 1–93. https://doi.org/10.1007/s10462-022-10286-2

22. Chataut, R., & Akl, R. (2020). Massive MIMO systems for 5G and beyond networks—Overview, recent trends, challenges, and future research direction. *Sensors, 20*, 2753. https://doi.org/10.3390/s20102753

23. Dutt, S., Ahuja, N. J., & Kumar, M. (2021). An intelligent tutoring system architecture based on fuzzy neural network (FNN) for special education of learning disabled learners. *Education and Information Technologies, 27*, 2613. https://doi.org/10.1007/s10639-021-10713-x

24. Ahuja, N. J., Dutt, S., Choudhary, S. L., & Kumar, M. (2022). Intelligent tutoring system in education for disabled learners using human–computer interaction and augmented reality. *International Journal of Human–Computer Interaction*, 1–13. https://doi.org/10.1080/10447318.2022.2124359

25. Trakadas, P., Simoens, P., Gkonis, P., Sarakis, L., Angelopoulos, A., Ramallo-González, A. P., Skarmeta, A., Trochoutsos, C., Calvo, D., Pariente, T., Chintamani, K., Fernandez, I., Irigaray, A. A., Parreira, J. X., Petrali, P., Leligou, N., & Karkazis, P. (2020). An artificial intelligence-based collaboration approach in industrial IoT manufacturing: Key concepts, architectural extensions and potential applications. *Sensors, 20*, 5480. https://doi.org/10.3390/s20195480

26. Chaudhry, N., & Yousaf, M. M. (2018). Consensus algorithms in blockchain: Comparative analysis, challenges and opportunities. In *2018 12th international conference on open source systems and technologies (ICOSST)* (pp. 54–63). https://doi.org/10.1109/ICOSST.2018.8632190

27. Wieringa, J., Kannan, P. K., Ma, X., Reutterer, T., Risselada, H., & Skiera, B. (2019). Data analytics in a privacy-concerned world. *Journal of Business Research, 122*, 915. https://doi.org/10.1016/j.jbusres.2019.05.005

28. Gatteschi, V., Lamberti, F., Demartini, C., Pranteda, C., & Santamaría, V. (2018). Blockchain and smart contracts for insurance: Is the technology mature enough? *Future Internet, 10*, 20. https://doi.org/10.3390/fi10020020

29. Treiblmaier, H., & Sillaber, C. (2021). The impact of blockchain on e-commerce: A framework for salient research topics. *Electronic Commerce Research and Applications, 48*, 101054. https://doi.org/10.1016/j.elerap.2021.101054

30. Lochau, M., Mennicke, S., Baller, H., & Ribbeck, L. (2016). Incremental model checking of delta-oriented software product lines. *Journal of Logical and Algebraic Methods in Programming, 85*(1), 245–267. https://doi.org/10.1016/j.jlamp.2015.09.004

31. Dingman, W., Cohen, A., Ferrara, N., Lynch, A., Jasinski, P., Black, P., & Deng, L. (2019). Defects and vulnerabilities in smart contracts, a classification using the NIST bugs framework. *International Journal of Networked and Distributed Computing., 7*, 121. https://doi.org/10.2991/ijndc.k.190710.003

32. Ducrée, J., Etzrodt, M., Bartling, S., Walshe, R., Harrington, T., Wittek, N., Posth, S., Wittek, K., Ionita, A., Prinz, W., Kogias, D., Paixão, T., Peterfi, I., & Lawton, J. (2021). Unchaining

collective intelligence for science, research, and technology development by blockchain-boosted community participation. *Frontiers in Blockchain, 4,* 631648. https://doi.org/10.3389/fbloc.2021.631648

33. Sedlmeir, J., Lautenschlager, J., Fridgen, G., et al. (2022). The transparency challenge of blockchain in organizations. *Electron Markets, 32,* 1779–1794. https://doi.org/10.1007/s12525-022-00536-0

34. Reyna, A., Martín, C., Chen, J., Soler, E., & Díaz, M. (2018). On blockchain and its integration with IoT. Challenges and opportunities. *Future Generation Computer Systems, 88,* 173–190. https://doi.org/10.1016/j.future.2018.05.046

35. Montanaro, A. (2016). Quantum algorithms: An overview. *npj Quantum Information, 2,* 15023. https://doi.org/10.1038/npjqi.2015.23

36. Leible, S., Schlager, S., Schubotz, M., & Gipp, B. (2019). A review on blockchain technology and blockchain projects fostering open science. *Frontiers in Blockchain, 2,* 486595. https://doi.org/10.3389/fbloc.2019.00016

37. Wang, W., Yu, Y., & Du, L. (2022). Quantum blockchain based on asymmetric quantum encryption and a stake vote consensus algorithm. *Scientific Reports, 12,* 8606. https://doi.org/10.1038/s41598-022-12412-0

38. Andoni, M., Robu, V., Flynn, D., Abram, S., Geach, D., Jenkins, D., & Peacock, A. (2019). Blockchain technology in the energy sector: A systematic review of challenges and opportunities. *Renewable and Sustainable Energy Reviews, 100,* 143–174. https://doi.org/10.1016/j.rser.2018.10.014

39. Casino, F., Dasaklis, T. K., & Patsakis, C. (2018). A systematic literature review of blockchain-based applications: Current status, classification and open issues. *Telematics and Informatics, 36,* 55. https://doi.org/10.1016/j.tele.2018.11.006

40. Jyothi Ahuja, N., & Dutt, S. (2022). Implications of quantum science on industry 4.0: Challenges and opportunities. https://doi.org/10.1007/978-3-031-04613-1_6.